MICRO IRRIGATION SCHEDULING AND PRACTICES

Innovations and Challenges in Micro Irrigation

MICRO IRRIGATION SCHEDULING AND PRACTICES

Edited by
Megh R. Goyal, PhD, PE
Balram Panigrahi, PhD
Sudhindra N. Panda, PhD

Apple Academic Press, Inc.
3333 Mistwell Crescent
Oakville, ON L6L 0A2 Canada

Apple Academic Press, Inc.
9 Spinnaker Way
Waretown, NJ 08758 USA

© 2017 by Apple Academic Press, Inc.

First issued in paperback 2021

Exclusive worldwide distribution by CRC Press, a member of Taylor & Francis Group
No claim to original U.S. Government works

ISBN 13: 978-1-77-463659-6 (pbk)
ISBN 13: 978-1-77-188552-2 (hbk)

Library and Archives Canada Cataloguing in Publication

Micro irrigation scheduling and practices / edited by Megh R. Goyal, PhD, PE, Balram Panigrahi, PhD, Sudhindra N. Panda, PhD.
(Innovations and challenges in micro irrigation ; volume 7)
Includes bibliographical references and index.
Issued in print and electronic formats.
ISBN 978-1-77188-552-2 (hardcover).--ISBN 978-1-315-20738-4 (PDF)
1. Microirrigation--Management. I. Goyal, Megh Raj, editor II. Panigrahi, Balram, author, editor III. Panda, Sudhindra N., editor V. Series: Innovations and challenges in micro irrigation v. ; 7

S619.T74M525 2017   631.5'87   C2017-900753-X   C2017-900754-8

Library of Congress Cataloging-in-Publication Data

Names: Goyal, Megh Raj, editor. | Panigrahi, Balram, editor. | Panda, Sudhindra N., editor.
Title: Micro irrigation scheduling and practices / editors: Megh R. Goyal, Balram Panigrahi, Sudhindra N. Panda.
Other titles: Innovations and challenges in micro irrigation ; [v. 7] Description: Waretown, NJ : Apple Academic Press, 2017. | Series: Innovations and challenges in micro irrigation ; [volume 7] | Includes bibliographical references and index.
Identifiers: LCCN 2017002978 (print) | LCCN 2017005255 (ebook) | ISBN 9781771885522 (hardcover : alk. paper) | ISBN 9781315207384 (ebook)
Subjects: LCSH: Microirrigation. | Irrigation scheduling.
Classification: LCC S619.T74 M526 2017 (print) | LCC S619.T74 (ebook) | DDC 31.5/87--dc23
LC record available at https://lccn.loc.gov/2017002978

Apple Academic Press also publishes its books in a variety of electronic formats. Some content that appears in print may not be available in electronic format. For information about Apple Academic Press products, visit our website at **www.appleacademicpress.com** and the CRC Press website at **www.crcpress.com**

CONTENTS

LIST OF CONTRIBUTORS

M. Anantachar
Professor and Head, FMPE Department, College of Agricultural Engineering, University of Agricultural Sciences, Raichur – 584101, India, Mobile: +91-9480163906; E-mail: anantachar3@gmai.com

M. S. Ayyanagowder
Professor, Soil and Water Engineering Department, College of Agricultural Engineering, University of Agricultural Sciences, Raichur – 584101, India, Mobile: +91-9448001138; E-mail: msaswe@gmail.com

Arpna Bajpai
Former MTech Student, Department of Soil and Water Engineering, College of Agricultural Engineering, Jabalpur 482004, MP, India, Mobile: +91-8989428391; E-mail: arpnabajpai@gmail.com

S. R. Balanagoudar
Assistant Professor, SAC Department, College of Agricultural Engineering, University of Agricultural Sciences, Raichur – 584101, India, Mobile: +91-9845013517; E-mail: balanagoudar@yahoo.com

Anuradha Behera
Former MTech Student, Department of Soil and Water Conservation Engineering, College of Agricultural Engineering and Technology, Orissa University of Agriculture and Technology, Bhubaneswar – 751003, Odisha, India, Mobile: +91-9234207011; E-mail: radhambsr17@gmail.com

V. A. Bhadane
Research Scholar, Department of Irrigation and Drainage Engineering, Dr. Panjabrao Deshmukh Krishi Vidyapeeth, Akola – 444104 (Maharashtra), India, Mobile: +91-7385071199; E-mail: vinitbhadane2310@gmail.com

P. K. Bora
Associate Professor, College of Agricultural Engineering and Postharvest Technology, Ranipool, Gangtok – 737135, Sikkim, India, E-mail: pradip66@gmail.com

Hillolmoy Chakraborty
PhD Research Scholar (Water Resources Development and Management), Department of Agricultural Engineering, Triguna Sen School of Technology, Assam University (Central University), Silchar – 788011, Cachar, Assam, India, Mobile: +91-9435588268, E-mail: hillolmoychakraborty@gmail.com

S. V. Chavan
Assistant Professor, SIT – Kolhapur, Department of Soil and Water Conservation Engineering, College of Agricultural Jainapur, Kolhapur – 416101, India, Mobile: +91-9975322017; E-mail: chavan664@gmail.com

Kishor Choudhari
Design and Project Engineer, Jain Irrigation System Ltd., Jain Plastic Park, N.H. No. 6, BAMBHORI, Jalgaon – 425001, Maharashtra, India; Tel.: +91-2572258011; Mobile: +91-9767778468; E-mail: kishorchoudhari19@gmail.com

Manish Debnath
PhD Research Scholar (Soil and Water Conservation Engineering), ICAR-Indian Agricultural Research Institute (IARI), New Delhi – 110012, India, Mobile: +91-8750761616; E-mail: debnathmanish55@gmail.com

M. M. Deshmukh
Associate Professor, Department of Irrigation and Drainage Engineering, Dr. Panjabrao Deshmukh Krishi Vidyapeeth, Akola – 444104 (Maharashtra), India, Mobile: +91-9921130260; E-mail: mahendradeshmukh@yahoo.com

S. M. Ghawade
Breeder cum Horticulturist, Chili and Vegetable Unit, Dr. Panjabrao Deshmukh Krishi Vidyapeeth, Akola – 444104 (Maharashtra), India, Mobile: +91-9657725844; E-mail: smghawade@gmail.com

Megh R. Goyal
Retired Professor in Agricultural and Biomedical Engineering, University of Puerto Rico – Mayaguez Campus; and Senior Technical Editor-in-Chief in Agriculture Sciences and Biomedical Engineering, Apple Academic Press, Inc., PO Box 86, Rincon – PR – 00677 – USA. E-mail: goyalmegh@gmail.com

Rajendra Gupta
Scientist, Indian Institute of Sugarcane Research, Raibareli Road, P.O. Dilkusha, Lucknow – 226002, UP, India. Tel.: +91-522-2480726; E-mail: iisrlko@sancharnet.in; gupta_iisr@mail.com

S. K. Gupta
INAE Distinguished Professor and Former Project Coordinator, AICRP on SAS & USWA, Central Soil Salinity Research Institute, Karnal – 1131001, India, Mobile: +91-9416081613; E-mail: s.gupta@icar.gov.in

A. Hugar
Professor, Department of Horticulture, ACR – Raichur, University of Agricultural Sciences, Raichur – 584104, Karnataka, India, Mobile: +91-9448757567; E-mail: ashok2_5_62@yahoo.co.in

A. B. Joshi
Manager, Department of Product development, Jain Irrigation systems Ltd, Jalgaon, Maharashtra, India, Mobile: +91-942283402; E-mail: abhijitjoshi@jains.com

M. U. Kale
Assistant Professor, Department of Irrigation and Drainage Engineering, Dr. Panjabrao Deshmukh Krishi Vidyapeeth, Akola – 444104 (Maharashtra), India, Mobile: +91-9422767788; E-mail: kale921@gmail.com

U. M. Khodke
Associate Dean and Principal, College of Agricultural Engineering and Technology; Head of Dept. of Agri. Eng.; Vasantrao Naik Marathwada Krishi Vidyapeeth, Parbhani – 431402, India, Mobile: +91-9422178025 Tel.: +91-2452–223230; E-mail: umkhodke@rediffmail.com

Dinaranjan Mahapatra
PhD Research Scholar, Department of Soil and Water Conservation Engineering, College of Agricultural Engineering and Technology, Orissa University of Agriculture and Technology (OUAT), Bhubaneswar – 751003, Odisha, India, Mobile: +91- 9439205010; E-mail: dinaranjan@gmail.com

Anil Kumar Mishra
Principal Scientist, Water Technology Centre, ICAR-Indian Agricultural Research Institute (IARI), New Delhi – 110012, India, Mobile: +91-9868846577; E-mail: dranilkumarmishra1@gmail.com

Kirti Sundar Mohanty
Former MTech Student, Department of Soil and Water Conservation Engineering, College of Agricultural Engineering and Technology, Orissa University of Agricultural & Technology, Bhubaneswar – 751003, Odisha, India. Current Address: Kirti Group, Kacaramal, Nakara, Cuttack – 754001; Mobile: +91-9937746130; E-mail: agrokirtigroup@gmail.com

Sheelabhadra Mohanty
Senior Scientist, ICAR-Indian Institute of Water Management, Opp. Rail Vihar, Chandrasekharpur, Bhubaneswar, Odisha – 751023, India, Mobile: 91–9438008253; E-mail: smohanty_wtcer@yahoo.co.in; smohanty.wtcer@gmail.com

A. K. Nayak
Scientist, Irrigation and Drainage Engineering Division, ICAR-Central Institute of Agricultural Engineering, Bhopal 462038, MP, India, Mobile: +91-8989206421; E-mail: anayak62@gmail.com

M. Nemichandrappa
Professor, Soil and Water Engineering Department, College of Agricultural Engineering, University of Agricultural Sciences, Raichur – 584101, India, Mobile: +91-9448303255; E-mail: Nemichandrappa@gmail.com

Sudhindra N. Panda
Director, National Institute of Technical Teachers Training and Research (NITTTR), [Ministry of Human Resources Development, Govt. of India], CSIR Road, Taramani, Chennai, Tamil Nadu 600113, India, Tel.: +91-44–22541982; E-mail: dir@nitttrc.ac.in; sudhindra.n.panda@gmail.com; Website: www.nitttrc.ac.in/

Balram Panigrahi
Professor and Head, Department of Soil and Water Conservation Engineering, College of Agricultural Engineering and Technology, Orissa University of Agriculture and Technology (OUAT), Bhubaneswar – 751003, Odisha, India, Mobile: +91-9437882699; E-mail: kajal_bp@yahoo.co.in

Neelam Patel
Principal Scientist, Water Technology Centre, ICAR-Indian Agricultural Research Institute (IARI), New Delhi – 110012, India, Mobile: +91-9868060359; E-mail: np_wtc@yahoo.com

M. G. Patil
Professor, Horticulture Department, College of Agricultural Engineering, University of Agricultural Sciences, Raichur – 584101, India, Mobile: +91-9741855127; E-mail: dr.patil_mg@rediffmail.com

Saroj Kumar Pattanaaik
Assistant Professor (Senior Scale), College of Horticulture & Forestry, Central Agricultural University, Pasighat, Arunachal Pradesh – 791102, India, Mobile: +91-9436630596; E-mail: saroj_swce@rediffmail.com

J. C. Paul
Associate Professor, Department of Soil and Water Conservation Engineering, College of Agricultural Engineering and Technology, Orissa University of Agricultural and Technology, Bhubaneswar – 751003, Odisha, India, Mobile: +91-9437762584; E-mail: jcpaul66@gmail.com

B. S. Polisgowdar
Professor, Department of Soil and Water Engineering, College of Agril. Engineering, University of Agricultural Sciences, Raichur – 584104, Karnataka, India, Mobile: +91-9448570701; E-mail: polisgowdar61@yahoo.com

S. K. Pyasi
Principal Scientist, Department of Soil and Water Engineering, College of Agricultural Engineering, Adhartal, Jabalpur 482004, MP, India, Mobile: +91-9301320977; E-mail: skpyasi@gmail.com

Lala I. P. Ray
Assistant Professor (Water Resource Management), School of Natural Resource Management, College of Postgraduate Studies (Central Agricultural University at Imphal), Umiam, Barapani – 793103, Meghalaya, India. Tel.: +91-364–2570031/2570614; Fax: +91-364–2570030; Mobile: +91-9436336021; E-mail: lalaipray@rediffmail.com

Mallikarjun Reddy
PhD Student, Soil and Water Engineering Department, College of Agricultural Engineering, University of Agricultural Sciences, Raichur – 584101 India, Mobile: +91-8861433579; E-mail: mallureddycae2026@gmail.com

N. Sahoo
Associate Professor, Department of Soil and Water Conservation Engineering, College of Agricultural Engineering and Technology, Orissa University of Agriculture and Technology (OUAT), Bhubaneswar – 751003, Odisha, India, Mobile: +91- 9437191308; E-mail: narayan_swce@yahoo.co.in

U. Satishkumar
Professor, Department of Soil and Water Engineering, College of Agricultural Engineering, University of Agricultural Sciences, Raichur – 584104, Karnataka, India, Mobile: +91-9448973765; E-mail: uskrcrcae01@gmail.com

C. K. Saxena
Senior Scientist, Irrigation and Drainage Engineering Division, ICAR-Central Institute of Agricultural Engineering, Bhopal 462038, MP, India. Tel.: +91-7552521152; Fax: +91-755–2734016; Mobile: +91-9407554272; E-mail: cksaxena@gmail.com, Chandra.Saxena@icar.gov.in

Laxmi Narayan Sethi
Associate Professor, Department of Agricultural Engineering, Triguna Sen School of Technology, Assam University (Central University), Silchar – 788011, Cachar, Assam, India, Mobile: +91-9864372058, 9401847943; E- mail: lnsethi06@gmail.com

K. Siangshai
Project Assistant, DST (Water Technology Initiative) Project, College of Postgraduate Studies, Umiam – 793103, Meghalaya, India, E-mail: kesparsiangsh@gmail.com

Ayyanna D. Siddapur
PhD Research Scholar, Department of Soil and Water Engineering, College of Agricultural Engineering, University of Agricultural Sciences, Raichur – 584104, Karnataka, India, Mobile: +91-8105130846; E-mail: ayyasiddapur@gmail.com

A. K. Singh
Assistant Professor, School of Natural Resource Management, College of Postgraduate Studies, Umiam – 793103, Meghalaya, India, E-mail: adityakumar1972@yahoo.co.in

Ram Singh
Associate Professor, School of Social Sciences, College of Postgraduate Studies, Umiam – 793103, Meghalaya, India, E-mail: ramsingh.cau@gmail.com

Ramadhar Singh
Principal Scientist, Irrigation and Drainage Engineering Division, ICAR-Central Institute of Agricultural Engineering, Bhopal 462038, MP, India, Mobile: +91-9685636309; E-mail: rsingh067bpl@gmail.com

R. C. Srivastava
Vice-Chancellor, Rajendra Agricultural University, Pusa, Samastipur, Bihar – 848125, India. Tel.: +91-6274–240226; Mobile: +91-9040033323; E-mail: ramesh_cari@yahoo.co.in, vcraupusa@gmail.com

Ch. R. Subudhi
Associate Professor, Department of Soil and Water Conservation Engineering, College of Agricultural Engineering and Technology, Orissa University of Agriculture & Technology, Bhubaneswar – 751003, Odisha, India, Mobile: 91–9437645234; E-mail: rsubudhi5909@gmail.com

I. Suting
Junior Research Fellow, DST (Water Technology Initiative) Project, College of Postgraduate Studies, Umiam – 793103, Meghalaya, India, E-mail: iba.suting@rediffmail.com

S. B. Wadatkar
Professor and Head, Department of Irrigation and Drainage Engineering, Dr. Panjabrao Deshmukh Krishi Vidyapeeth, Akola – 444104 (Maharashtra), India, Mobile: +91-9423129093; E-mail: wadatkarsb@rediffmail.com

V. B. Wali
Professor, Department of Statistics, University of Agricultural Sciences, Raichur – 584104, Karnataka, India, Mobile: +91-9901500145; E-mail: vbw06@rediffmail.com

LIST OF ABBREVIATIONS

AICRP	All India Co-ordinated Research Project
APEDA	Agricultural & Processed Food Products Export Development Authority
ARV	air release valve
ASCE	America Society of Civil Engineers
ASM	available soil moisture content
BCE	Before the Common Era
CCU	Christiansen's coefficient of uniformity
CD	critical difference
CEFV	coefficient for emitter flow variation
CFD	computational fluid dynamics
CPE	cumulative pan evaporation
CV	coefficient of variation
CWPF	crop water production functions
DAP	days after planting
DAS	days after sowing
DC	direct current
DI	duration of irrigation
DIS	drip irrigation system
DU	double union
EC	electrical conductivity
EGL	energy gradient line
ET	evapotranspiration
et al.	et alibi
FDR	frequency domain reflectrometry
FYM	farm yard manure
GDIS	gravity-fed drip irrigation system
GI	galvanized iron
GP	galvanized pipe
HDPE	high density polyethylene
HP	horse power

IC	integrated circuit
ICAR	Indian Council of Agricultural Research
IDE	Integrated Development Environment
IIT	Indian Institute of Technology
IMD	Indian Meteorological Department
INAE	Indian National Academy of Engineering
ISAE	Indian Society of Agricultural Engineers
IW	irrigation depth
LAI	leaf area index
LCD	liquid crystal display
LDPE	low-density polyethylene
LED	light emitting diodes
LLDPE	linear low density polyethylene
LM	monolithic
MAD	management allowable deficit
MIS	micro irrigation system
MP	Madhya Pradesh
MSL	mean sea level
MT	million ton
N	nitrogen
NER	north eastern region
NRV	non return valve
OC	organic carbon
OOP	object oriented programming
OUAT	Orissa University of Agriculture and Technology
P	phosphorous
PE	maximum pan evaporation
PM	polyethylene mulch
ppm	parts per million
PVC	poly vinyl chloride
PWR	peak water requirement
RDF	recommended dose of fertilizer
REGL	revised energy gradient line
RTC	real time clock
SAS&USWA	salt affected soil and use of saline water in agriculture
SCU	statistical coefficient of uniformity

SDI	sub-surface drip irrigation
SDU_{lq}	low quarter distribution uniformity
SMW	Standard Meteorological Week
SRI	System of Rice Intensification
SSDI	Subsurface drip irrigation
TDR	time domain reflectrometry
TSS	total soluble salts
TTT	temperature-time-threshold
U.S.A.	United States of America
UK	United Kingdom
US$	United States Dollar
UV	ultra violet
VR	variable resistors
WR	water requirement
WRC	crop water requirement for the growing period
WSCU	Wilcox–Swailes coefficient of uniformity
WUE	water use efficiency

LIST OF SYMBOLS

$\Delta H_{lateral}$	head loss in lateral
ΔH_{main}	head loss in main
$\Delta H_{submain}$	friction loss in submain
η_{motor}	motor efficiency
η_{pump}	pump efficiency
μ	dynamic viscosity of water
ρ	density of water
Φ	diameter
%	percentage
A	area allocated to each plant
A	area of the crop
A	daily evapotranspiration
a	efficiency of motor
A_2	cross sectional area of throat
Ac	plant canopy area
A_p	effective area irrigated
B	canopy factor
B	crop factor (depends on growth stage and foliage caver)
b	efficiency of pump
B	inlet diameter
B:C	Benefit:Cost
C	canopy factor
C	crop coefficient
C	Hazen–William constant (140 for PVC pipe)
cc	crop-to-crop spacing
Cd	coefficient of discharge
cm	centimeter
D	crop area (row-to-row spacing in m crop-to-crop spacing)
d	diameter of pipe
D	elevation difference
D	internal diameter of pipe, cm

d	throat diameter
df	degree of freedom
$D_{lateral}$	inside diameter of the lateral
$D_{submain}$	inside diameter of the submain
E	efficiency of drip irrigation
E	efficiency of irrigation system
E	emission uniformity of drip irrigation system
e	total number of emitters
Ea	water application efficiency
Ep	pan evaporation
ET_0	reference crop evapotranspiration
ETc	crop evapotranspiration
E_u	emission uniformity of drip system
Eu	water use efficiency
f	friction factor
F	outlet factor
F	reduction factor due to multiple openings in pipe
Fx	mass force in x direction
Fy	mass force in y direction
Fz	mass force in z direction
g	acceleration due to gravity
g	gram
Ge	ground water contribution
h	operating pressure head of drip system
H	total head
ha	hectares
H_d	delivery head
H_E	head loss due to elevation difference
H_e	operating pressure of emitters
H_f	head loss due to friction
$H_f (100)$	head loss due to friction per 100 meter of pipe length
$H_{lateral}$	friction loss in lateral
Hp	power of the pump
H_s	suction head
I_n	net irrigation requirement of the crop
K	constant
K	potassium

K_c	crop factor or crop coefficient
Kg	kilogram
kg/cm²	kilogram per centimeter square
kgf cm⁻²	Kg force per square centimeter
Kp	pan coefficient
kPa	Kilo Pascal
L	length of the land
L_{ateral}	flow rate in the lateral
lh⁻¹	liter per hour
lit/m/day	liter per meter per day
$L_{lateral}$	length of lateral
$L_{lateral1}$	length of lateral 1
$L_{lateral2}$	length of lateral 2
lph	liter per hour
lph/day	liter per hour per day
lps	liter per second
$L_{sub\ main}$	length of submain
m	meter
M ha	million hectare
m³/h	meter cube per hour
mg	milligram
mg/l	milligram per liter
Mha	million hectare
micron	10⁻⁶ meter
mm	millimeter
mm/day	millimeter per day
n	number of dripper required per tree
N_1	total number of lateral on sub main
Nm	nano meter
P	precipitation
p	pressure of fluid tiny body
P_1	inlet pressure
P_2	throat pressure
pH	power of hydrogen
Q	flow of water in pipe, lps
q	peak water requirement
Q	quantity of water applied

q	Quintal, 1.00 q = 100 kg
Q_1	discharge rate of one lateral
Q_{act}	actual motive flow
Q_{cfd}	actual motive flow calculated by computation fluid dynamics
$Q_{lateral}$	discharge rate of one lateral
q_{min}	minimum emitter flow rate
$Q_{submain}$	flow rate in the sub main
Q_{the}	theoretical motive flow
R	surface runoff
R_e	effective rainfall
rr	row-to-row spacing
Rs.	Indian Rupees (1 US$ = 60.00 Rs.)
T	total irrigation time
T	treatment
t ha^{-1}	ton per hectare
t(s)	time
T_p	total number of plant
u	velocity at x direction
U	velocity vector
V	net depth of irrigation
V	total volume of water applied
v	velocity at y direction
v	velocity of water in pipe
w	velocity at z direction
w	wetted area of the crop
W	wetting fraction
W	width of the land
W_b	soil moisture contribution
W_f	water delivered to the field
W_s	water stored in the root zone
X_1, X_2, X_3	root length at active tillering stage, flowering stage, and maturity stage, respectively
X_4	panicle length
X_5	test weight
Y	yield of the crop

PREFACE 1 BY MEGH R. GOYAL

Clogging is a serious chronic cause of failure of micro irrigation systems.
However if the irrigator uses the tools that are available, he can live a full productive and joyful life.
Giving back is very important to me, as it defines who I am.
I am an ordinary irrigation expert, as I still live like the reader.
I just can't see you, but I can enjoy that you have read my books on micro irrigation.
God bless you as you browse through my books that have been prepared for you only.
I can assure you that drip irrigation can potentially provide high application efficiency and application uniformity.
—Megh R. Goyal, Drip Man

This will naturally reduce the speed of your mental thoughts and then help to … dams, the same flooded water becomes the source of energy generation and *irrigation*. In the same way, when we channelize our *positive thought* pattern we get …
—B. K. Chandra Shekhar

During March 13 through March 17 of 2016, I along with my wife visited Jaipur (dʒaɪpuər; Devanāgarī: जयपुर; Nickname: The Pink City), capital city of Rajasthan in India. My eyes were widened to observe that micro irrigation is in operation at most of the historic sites of this city. Following information is available at https://en.wikipedia.org/wiki/Jaipur.

The city of Jaipur was founded in 18 November 1727 by Jai Singh II, the Raja of Amer who ruled from 1688 to 1758. He planned to shift his capital from Amer, 11 km from Jaipur to accommodate the growing population and increasing scarcity of water. Jai Singh consulted several books on architecture and architects while planning the layout of Jaipur. Under the architectural guidance of Vidyadhar Bhattacharya,

Jaipur was planned. The construction of the city began in 1727 and took four years to complete the major roads, offices and palaces. The city was divided into nine blocks, two of which contained the state buildings and palaces, with the remaining seven allotted to the public. Huge ramparts were built, pierced by seven fortified gates.

During the rule of Sawai Ram Singh, the city was painted pink to welcome the Prince of Wales, later Edward VII, in 1876. Many of the avenues remained painted in pink, giving Jaipur a distinctive appearance and the epithet Pink city. In the 19th century, the city grew rapidly and by 1900 it had a population of 160,000. The wide boulevards were paved and its chief industries were the working of metals and marble, fostered by a school of art founded in 1868. The city had three colleges, including a Sanskrit college (1865) and a girls' school (1867) opened during the reign of the Maharaja Ram Singh II.

In 2011, the city had a population of 3.1 million, making it the tenth most populous city in the country. Located at a distance of 260 km from the Indian capital New Delhi, it forms a part of the Golden Triangle tourist circuit along with Agra (240 km). Jaipur is a popular tourist destination in India and serves as a gateway to other tourist destinations in Rajasthan such as Jodhpur (348 km), Jaisalmer (571 km) and Udaipur (421 km)...

Jaipur Exhibition and Convention Centre (JECC) is Rajasthan's biggest convention and exhibition center. Visitor attractions in Jaipur include: the Hawa Mahal, Jal Mahal, City Palace, Amer Fort, Jantar Man-

tar, Nahargarh Fort, Jaigarh Fort, Galtaji, Govind Dev Ji Temple, Garh Ganesh Temple, Sri Kali Temple, Birla Mandir, Sanganeri Gate and the Jaipur Zoo. The Jantar Mantar observatory and Amer Fort are one of the World Heritage Sites. Hawa Mahal is a five-story pyramidal shaped monument with 953 windows that rises 15 meters (50 feet) from its high base. Sisodiya Rani Bagh and Kanak Vrindavan are the major parks in Jaipur.

Readers might wonder what these historic sites have to do with micro irrigation. In 1727, Jai Singh II was aware of problems of water scarcity, and he hired an architect (may be irrigation engineers were not available in those days) to cope with this rampant problem. The city water supply in Jaipur is one of the best in India. Water scarcity problems continue to cause headache to the city planners even today. Let us all join hands together to plan intelligent use of this rich resource of water.

My vision for micro irrigation technology has expanded globally. I am astonished to observe how this is expanding to tourist regions and especially to archeological sites with the number of visitors exceeding one million per year. Although no emphasis is made to draw attention of visitors to this valuable technology, yet there is a potential audience.

This book volume, *Micro Irrigation Scheduling and Practices*, presents: performance of vegetable, fruit and row crops; and practices in drip Irrigation design. The mission of this book volume is to serve as a reference manual for graduate and undergraduate students of agricultural, biological and civil engineering, horticulture, soil science, crop science and agronomy. I hope that it will be a valuable reference for professionals that work with micro irrigation and water management; and for professional training institutes, technical agricultural centers, irrigation centers, agricultural extension services, and other agencies that work with micro irrigation programs. I cannot guarantee the information in this book series will be enough for all situations.

After my first textbook, *Drip/Trickle or Micro Irrigation Management* by Apple Academic Press, Inc., and response from international readers, Apple Academic Press, Inc. has published for the world community the 10-volume series on *Research Advances in Sustainable Micro Irrigation*, edited by M. R. Goyal. The website http://appleacademicpress.com gives details on these ten book volumes. This book volume is one of the future volumes under book series, *Innovations and Challenges in Micro Irrigation*. Both book series are musts for those interested in irrigation planning and management, namely researchers, scientists, educators and students.

The contributions by the cooperating authors to this book series have been most valuable in the compilation of this volume. Their names are mentioned in each chapter and in the list of contributors. This book would not have been written without the valuable cooperation of Dr. Balram Panigrahi and Dr. Sudhindra N. Panda (both co-editors of this book) and these investigators, many of whom are renowned scientists who have worked in the field of micro irrigation throughout their professional careers.

I would like to thank editorial staff, Sandy Jones Sickels, Vice President, and Ashish Kumar, Publisher and President at Apple Academic Press, Inc., for making every effort to publish the book when the diminishing water resources are a major issue worldwide. Special thanks are due to the AAP Production staff for the quality production of this book.

We request the reader to offer us your constructive suggestions that may help to improve the next edition. The reader can order a copy of this book for the library, the institute or for a gift from http://appleacademic-press.com.

I express my deep admiration to my wife, Subhadra Devi Goyal, for understanding and collaboration during the preparation of this book. As an educator, there is a piece of advice to one and all in the world: *"Permit that our almighty God, our Creator, excellent Teacher and Micro Irrigation Designer, irrigate our life with His Grace of rain trickle by trickle, because our life must continue trickling on..."*

—*Megh R. Goyal, PhD, PE*
Senior Editor-in-Chief

PREFACE 2 BY BALRAM PANIGRAHI

Water is the most critical input for agriculture and plays a crucial role in maximizing production and productivity of crops. Since the demand of water in the non-agricultural sector is increasing day by day, its share for agriculture is decreasing at a faster rate. Its efficient utilization is basic to the survival of mankind and is highly essential for sustenance of agricultural production. It is necessary to economize the use of water for agriculture in order to bring more area under irrigation. Formulation of efficient and economically viable irrigation management strategies in order to irrigate more area with the existing limited water resources is the call of the day. Introduction of micro irrigation accelerates water saving and increases the water application efficiency up to 90%, thereby increasing the crop irrigated area, cropping intensity, production and productivity of crops, and consequently enhancing the socio-economic status of the farmers.

Innovations are essential for refinement and upgradation of existing technology in all fields, including micro irrigation. Although the micro irrigation technology has been popularized in many countries, there is not yet much documentation available, which needs to be spread to the farming community for its wider adoption. To provide a complete and comprehensive knowledge on micro irrigation, the authors have attempted to bring out this book, *Micro Irrigation Scheduling and Practices* by Apple Academic Press, Inc.

The book contains three parts with 16 chapters. Part I, entitled *Performance of Vegetable Crops,* contains four chapters; Part II, entitled *Performance of Fruit and Row Crops,* contains seven chapters; and Part III, entitled *Practices in Drip Irrigation Design,* contains five chapters. Micro irrigation scheduling and practices have been discussed in various chapters of the book for various fruit, row and vegetables crops and flowers, including capsicum, chili, watermelon, banana, kinnow, litchi, rice, sugarcane, sorghum, marigold, etc. In addition, the design principles of micro irrigation considering discharge, pressure variations and head loss are discussed. A software program for design of drip irrigation for multi crops is presented in this book.

The book will serve as an invaluable resource for graduate and undergraduate students in the field of agriculture, agricultural, biological and water resources engineering. The book will be helpful for all academicians, researchers, practicing engineers, agronomists, and extension personnel. The contributions by the authors of different chapters of this book are very valuable, and without their support this book would have not been published successfully. Their names are mentioned in each chapter and also separately in the list of contributors and are duly acknowledged.

I take the opportunity to offer my heartfelt obligations to *Prof. Megh R. Goyal, "Father of Irrigation Engineering of 20th Century in Puerto Rico"*, who has benevolently given me an opportunity to serve as an editor of the book. He has been instrumental in spreading this technology to communities involved in micro irrigation throughout the world. We all applaud his efforts. I am also thankful to *Prof. S. N. Panda* who has contributed to bring out this book in the present form as an editor. My special thanks to all the editorial staff of Apple Academic Press, Inc. for making every effort to publish the book.

I express my deep obligations to my family, friends, and colleagues for their help and moral support during preparation of the book. Readers are requested to offer constructive suggestions that may help to improve the next edition.

—***Balram Panigrahi, PhD***
Editor

PREFACE 3 BY
SUDHINDRA N. PANDA

Water is one of the most vital resources for sustainable development of agriculture. The resource is under severe stress due to industrialization, urbanization, changes in life styles, modern agricultural practices, climate change, and agricultural virtual water flow. Serious water shortages have been observed throughout the world, with an increase in competition for clean water, as water resources reach full exploitation. Sustainable management of water in a 'green' economy is necessary with an increase of irrigated areas in the future and diversion of fresh water for greater demand of industrial and domestic use. The major challenge is to increase food production with less water and/or with gray water with the help of different water-saving technologies and management methods.

Throughout the world, use of micro irrigation for improving water productivity and economizing its use has increased. It is one of the most efficient methods for providing water to the crops. Micro irrigation can help in gaining maximum economic profit under scarce water supply conditions and climate change. However, it requires specific knowledge of crop response to water as it varies significantly among crops and their growth stages. Scientific research studies provide information on advances in knowledge, technology, and applications.

The main aim of this book is to provide informative and comprehensive knowledge on micro irrigation scheduling and practices. The book incorporates the latest information on the subject and covers the area of performance, practices, and design. Chapters in the book provide in-depth knowledge and analyseis on various aspects of micro irrigation. It includes performance of vegetable, fruit and row crops for different scheduling and practices. Design aspects of micro irrigation system has also been discussed in the book. It will serve as a valuable reference and will assist students, academicians, researchers, water resources professionals, extensionists, farmers, and decision-makers in gaining knowledge on micro irrigation. I am greatly indebted to contributing authors of different chapters in the book.

—*Sudhindra N. Panda, PhD*
Editor

WARNING/DISCLAIMER

READ CAREFULLY

The goal of this compendium, *Micro Irrigation Scheduling and Practices,* is to guide the world engineering community on how to efficiently design for economical crop production. The reader must be aware that dedication, commitment, honesty, and sincerity are most important factors in a dynamic manner for a complete success. This is not a one-time reading of this compendium. Read and follow every time.

The editors, the contributing authors, the publisher, and the printer have made every effort to make this book as complete and as accurate as possible. However, there still may be grammatical errors or mistakes in the content or typography. Therefore, the contents in this book should be considered as a general guide and not a complete solution to address any specific situation in irrigation. For example, one size of irrigation pump does not fit all sizes of agricultural land and to all crops.

The editors, the contributing authors, the publisher, and the printer shall have neither liability nor responsibility to any person, any organization, or entity with respect to any loss or damage caused, or alleged to have caused, directly or indirectly, by information or advice contained in this book. Therefore, the purchaser/reader must assume full responsibility for the use of the book or the information therein.

The mention of commercial brands and trade names are only for technical purposes. A particular product is not endorsed over to another product or equipment not mentioned. The editors, the cooperating authors, the educational institutions, and the publisher Apple Academic Press, Inc., do not have any preference for a particular product.

All weblinks that are mentioned in this book were active on December 01, 2016. The editors, the contributing authors, the publisher, and the printing company shall have neither liability nor responsibility, if any of the weblinks is inactive at the time of reading of this book.

ABOUT THE SENIOR EDITOR-IN-CHIEF

Megh R. Goyal, PhD, PE
Retired Professor in Agricultural and Biomedical Engineering, University of Puerto Rico, Mayaguez Campus Senior Acquisitions Editor, Biomedical Engineering and Agricultural Science, Apple Academic Press, Inc.

Megh R. Goyal, PhD, PE, is a Retired Professor in Agricultural and Biomedical Engineering from the General Engineering Department in the College of Engineering at the University of Puerto Rico–Mayaguez Campus; and Senior Acquisitions Editor and Senior Technical Editor-in-Chief in Agriculture and Biomedical Engineering for Apple Academic Press, Inc.

He has worked as a Soil Conservation Inspector and as a Research Assistant at Haryana Agricultural University and Ohio State University. He was the first agricultural engineer to receive the professional license in Agricultural Engineering in 1986 from the College of Engineers and Surveyors of Puerto Rico. On September 16, 2005, he was proclaimed as "Father of Irrigation Engineering in Puerto Rico for the Twentieth Century" by the ASABE, Puerto Rico Section, for his pioneering work on micro irrigation, evapotranspiration, agroclimatology, and soil and water engineering. During his professional career of 45 years, he has received many prestigious awards, including Scientist of the Year, Blue Ribbon Extension Award, Research Paper Award, Nolan Mitchell Young Extension Worker Award, Agricultural Engineer of the Year, Citations by Mayors of Juana Diaz and Ponce, Membership Grand Prize for ASAE Campaign, Felix Castro Rodriguez Academic Excellence, Rashtrya Ratan Award and Bharat Excellence Award and Gold Medal, Domingo Marrero

Navarro Prize, Adopted son of Moca, Irrigation Protagonist of UPRM, Man of Drip Irrigation by Mayor of Municipalities of Mayaguez/Caguas/ Ponce, and Senate/Secretary of Agriculture of ELA, Puerto Rico.

The Water Technology Centre of Tamil Nadu Agricultural University in Coimbatore, India, has recognized Dr. Goyal as one of the experts "who rendered meritorious service for the development of micro irrigation sector in India" by bestowing the *"Award of Outstanding Contribution in Micro Irrigation."* This award was presented to Dr. Goyal during the inaugural session of the National Congress on "New Challenges and Advances in Sustainable Micro Irrigation" on March 1, 2017, held at Tamil Nadu Agricultural University.

A prolific author and editor, he has written more than 200 journal articles and textbooks and has edited over 48 books including: *Elements of Agroclimatology* (Spanish) by UNISARC, Colombia; and two bibliographies on drip irrigation.

He received his BSc degree in engineering from Punjab Agricultural University, Ludhiana, India; his MSc and PhD degrees from Ohio State University, Columbus; and his Master of Divinity degree from Puerto Rico Evangelical Seminary, Hato Rey, Puerto Rico, USA.

Apple Academic Press, Inc. (AAP) has published his books, namely, *Management of Drip/Trickle or Micro Irrigation,* and *Evapotranspiration: Principles and Applications for Water Management,* his ten-volume set on *Research Advances in Sustainable Micro Irrigation.* During 2016–2020, AAP will be publishing book volumes on emerging technologies/issues/ challenges under two book series, *Innovations and Challenges in Micro Irrigation,* and *Innovations in Agricultural and Biological Engineering.* Readers may contact him at: goyalmegh@gmail.com.

ABOUT THE EDITOR
DR. BALRAM PANIGRAHI

Balram Panigrahi, PhD
Professor and Head, Soil and Water Conservation Engineering, College of Agricultural Engineering and Technology, Orissa University of Agriculture and Technology, Bhubaneswar, India.

Dr. Balram Panigrahi is an agricultural engineer with specialization in soil and water engineering. Dr. Panigrahi is presently Professor and Head of the Department of Soil and Water Conservation Engineering (SWCE) at the College of Agricultural Engineering and Technology (CAET), Orissa University of Agriculture and Technology (OUAT), in Bhubaneswar, India. He also served as Chief Scientist of the Water Management Project and Associate Director of Research in the Regional Research Station of OUAT.

Dr. Panigrahi has published about 190 technical papers in different international and national journals and conference proceedings. He has written several book chapters, practical manuals, and monographs. He has also written two textbooks in the field of irrigation engineering. He has been awarded with 17 gold medals and awards, including the Jawaharlal Nehru Award for best postgraduate research in the field of natural resources management by the Indian Council of Agricultural Research, New Delhi; the Samanta Chandra Sekhar Award for best scientist in the state of Odisha, India; and the Gobinda Gupta Award as outstanding engineer of the state of Odisha, given by the Institution of Engineers (India), Odisha state center. He has also received a Japanese Master Fellowship for pursuing a master of engineering study at the Asian Institute of Technology, Thailand. In additional to being a reviewer and an editorial board member of several journals, he is the member of a number of professional societies at national and international levels. He has chaired several international and national conferences both in India and abroad. With 27 years of teaching and research experience, he has guided several PhD and many MTech students. Dr. Panigrahi's research interests include irrigation and drainage engineering, water management in rainfed and irrigated commands, and modeling of irrigation systems.

He obtained his BTech in agricultural engineering from Orissa University of Agriculture and Technology (OUAT), Bhubaneswar, Odisha, India, and his Master of Engineering in water resources engineering from the Asian Institute of Technology, Thailand. He was awarded a PhD in agricultural engineering from the Indian Institute of Technology, Kharagpur, India. Readers may contact him at: kajal_bp@yahoo.co.in

ABOUT THE EDITOR
DR. SUDHINDRA NATH PANDA

Sudhindra N. Panda, PhD
Director, National Institute of Technical Teachers Training and Research (NITTTR), Chennai, India

Dr. Sudhindra Nath Panda is currently Director of the National Institute of Technical Teachers Training and Research (NITTTR), Chennai, India. Earlier he was a Professor of Land and Water Resources Engineering at the Agricultural and Food Engineering Department of the Indian Institute of Technology (IIT), Kharagpur, India.

He has nearly 33 years of professional experience in teaching, research, extension, and training activities at several institutions, including IIT Kharagpur; the Orissa University of Agriculture and Technology, Bhubaneswar, India; the Water and Land Management Institute, Cuttack, India, and Punjab Agricultural University, Ludhiana, India, in several capacities, including Professor, Associate Professor, Reader (Engineering), Senior Scientist, and Assistant Professor. He was also a Visiting Researcher at the Division of Climatology and Water Resources, Arid Land Research Center, Tottori University, Tottori, Japan.

Prof. Panda is currently the Director (Soil and Water Engineering) of the Indian Society of Agricultural Engineers, New Delhi, India. He is an editorial board member of various journals and a member or fellow of many professional groups. He was the Founder and Head of the School of Water Resources, IIT Kharagpur, where he was instrumental in developing the course curriculum, basic infrastructure, laboratories, computational facilities, faculty recruitment, and creating job opportunities for MTech (water management) students. As a coordinator, Prof. Panda was instrumental in an arrangement of signing MOU between IIT Kharagpur and two German universities.

Prof. Panda is the recipient of various national and international awards and honors from various organizations and has handled almost a dozen international and national research projects. He has edited a workshop proceedings and written and published numerous research bulletins, book chapters, journal articles, conference presentations, and training manuals. He has chaired several international and national conferences both in India and abroad and has organized conferences and workshops. Panda is recognized nationally and internationally for his work in the field of systems approaches for integrated land and water resources planning and management and rainwater conservation and reuse for climate resilient agriculture. The field experimental setup at IIT Kharagpur on rainwater conservation and reuse technology, developed by Prof. Panda, has influenced farmers and NGOs of eastern India for its large-scale adoption in rainfed agriculture.

OTHER BOOKS ON MICRO IRRIGATION TECHNOLOGY BY APPLE ACADEMIC PRESS, INC.

Management of Drip/Trickle or Micro Irrigation
Megh R. Goyal, PhD, PE, Senior Editor-in-Chief

Evapotranspiration: Principles and Applications for Water Management
Megh R. Goyal, PhD, PE, and Eric W. Harmsen, Editors

Book Series: Research Advances in Sustainable Micro Irrigation
Senior Editor-in-Chief: Megh R. Goyal, PhD, PE
Volume 1: Sustainable Micro Irrigation: Principles and Practices
Volume 2: Sustainable Practices in Surface and Subsurface Micro Irrigation
Volume 3: Sustainable Micro Irrigation Management for Trees and Vines
Volume 4: Management, Performance, and Applications of Micro Irrigation Systems
Volume 5: Applications of Furrow and Micro Irrigation in Arid and Semi-Arid Regions
Volume 6: Best Management Practices for Drip Irrigated Crops
Volume 7: Closed Circuit Micro Irrigation Design: Theory and Applications
Volume 8: Wastewater Management for Irrigation: Principles and Practices
Volume 9: Water and Fertigation Management in Micro Irrigation
Volume 10: Innovation in Micro Irrigation Technology

Book Series: Innovations and Challenges in Micro Irrigation
Senior Editor-in-Chief: Megh R. Goyal, PhD, PE
Volume 1: Principles and Management of Clogging in Micro Irrigation
Volume 2: Sustainable Micro Irrigation Design Systems for Agricultural Crops: Methods and Practices
Volume 3: Performance Evaluation of Micro Irrigation Management: Principles and Practices

PART I

PERFORMANCE OF
VEGETABLE CROPS

CHAPTER 1

PERFORMANCE OF WINTER VEGETABLES UNDER GRAVITY-FED DRIP IRRIGATION SYSTEM

LALA I. P. RAY, I. SUTING, K. SIANGSHAI, A. K. SINGH, RAM SINGH, and P. K. BORA

CONTENTS

1.1   INTRODUCTION

Agriculture has been and continues to be the mainstay of many parts of world since antiquity and still remains as the highest consumer of water. Efficient utilization of freshwater in agricultural sector is sometimes not taken into consideration due to some reasons or other, both locally and globally. North Eastern Hilly (NEH) states of India have undulating topography with varied soil depth. Occurrence of predominantly sandy type acidic soil is also observed red in this NEH region. The agriculture in NEH region is mostly monsoon dependent with rice as main crop. However, maize based cropping system is found in the state of Sikkim. The cropping intensity of this region hardly remains between 120–156% except in the state of Assam. Due to unavailability of assured irrigation, farmers of this region used to go for only *kharif* crop mostly during monsoon season. The abundance of rainfall and untapped water resource potential of this region sometimes raised the question on this sorry figure of cropping intensity. With an average annual rainfall ranging from 2,480 to 6,350 mm, this NEH region is also endowed with rich water resource constituting 33% of country's water resources for hydropower, industrial and irrigation process, but its capacity still remains untapped. The NEH has the highest water availability of 16,500 cubic meters per capita and 44,180 cubic meters per hectare in the country. About 70% of the NEH region is dependent on agriculture for livelihood, yet it continues to be a net importer of food grains even for its own consumption excluding few pockets in Manipur, Assam and Tripura, the land productivity as compared to its potential is low [10].

The prevalent soil texture of NEH region is sandy, which has less water holding capacity; hence, water has to be applied slowly but more frequently to ensure a better crop yield. Along with feasible and effective water management practices through some water saving irrigation technologies like micro irrigation, the possibility of growing winter crop in this NEH region can be made possible. Drip irrigation system "one of the micro irrigation technologies" meets the requirement and has a major role to play in increasing cropping intensity in the hilly regions. It is a proven micro-irrigation technology, where productivity increases with judicious water usage. It requires as low as 0.1 kgf cm^{-2} (1 m) of head to operate. Since this system can operate with low head, source of water available at a relatively higher elevation can able to generate ample head to operate drip emitters; irrigation water can be applied by gravity using a drip irrigation setup [1, 6, 10]. The cost of pumping can be nullified taking into consideration the natural slope of the hilly terrain. It helps the farmers to go for winter crop during non-rainy seasons [3, 7, 14, 17].

The NEH region of India consists of hilly topography; and cropping areas are characterized by small plots. Although hand watering of crops in small plots can be followed, yet the application of drip irrigation system in these plots results in minimal water losses, significant labor reduction, with the potential to increase crop yield [14, 17, 18]. Growing vegetables during winter months using low cost irrigation system can save up to 40% water [6, 10]. This fact is highly relevant as the demand for vegetables is estimated to increase in the coming years [7, 14]. The important factor in drip irrigation is the ideal rate of water distribution along the soil profile. High rate water supply may cause deep percolation loss, whereas, very low rate may result in evaporation losses [9, 11–13, 16].

The major drawback of drip irrigation is high initial cost during installation of the system and operational cost incurred for its application. However, these drawbacks can be overcome through integration of gravity-fed component into the drip irrigation system [1, 10]. This system guarantees a simple cost efficient technology for providing irrigation facility to the cropping area generally ranging below 1 to 1.5 acre [3, 6].

In this chapter, an effort has been made to record the efficacy of Gravity-fed Drip Irrigation System (GDIS) to raise winter vegetables like Cabbage, Broccoli, Cauliflower and Baby corn. The crop performance as well as the system performance of GDIS was also evaluated.

1.2 MATERIALS AND METHODS

The study site (Barpani) is situated in the state of Meghalaya, India located at 91° 55' 25" east; 25° 41' 21" north at an altitude of 1,010 m above the mean sea level (MSL). The detailed description of the site is shown in Figure 1.1. This region normally experiences humid sub-tropical type of climate with high rainfall and cold winters. High relative humidity and low sunshine hours as compared to other parts of the India is also experienced in this pocket. The maximum temperature rises up to 30°C in the months of July–August and minimum falls down 5–6°C during the first week of January.

The weekly average of the maximum and minimum temperature during the cropping season ranged from 30.2–25.3°C and 20.6–12.1°C, respectively. The mean relative humidity ranged from 93.4–82.0% in the morning and 83.4–62.9% in the evening hours. The weekly average of

FIGURE 1.1   District map of Meghalaya [Source: http://ceomeghalaya.nic.in/]: District headquarter with a yellow circle.

Jaintia Hills Division	Khasi Hills Division	Garo Hills Division
West Jaintia Hills (Jowai)	East Khasi Hills (Shillong)	North Garo Hills (Resubelpara)
East Jaintia Hills (Khliehriat)	West Khasi Hills (Nongstoin)	East Garo Hills (Williamnagar)
	South West Khasi Hills (Mawkyrwat)	South Garo Hills (Baghmara)
	Ri-Bhoi (Nongpoh)	West Garo Hills (Tura)
		South West Garo Hills

maximum and minimum temperature was recorded as 30.2°C and 12.1°C during 32nd standard meteorological week (SMW) in the month of August and 43rd standard week in October, respectively.

1.2.1   RAINFALL AND EVAPORATION IN THE STUDY AREA

In this region, normally the monsoon season sets in during first fortnight of June and proceeds up to October however, the magnitude of rainfall decreases from September onwards. The average annual rainfall is 2,410.4 mm with 129 numbers of rainy days, with pre monsoon showers from February to May. The annual rainfall varied from 3,322.6 mm (1988) to 1,808.2 mm (1998) with an average monthly rainfall of 144.52, 283.68, 408.71, 439.02, 345.70 and 355.87 mm for the months of April, May, June, July, August and September, respectively. The highest rainfall of 242 mm was recorded in 38th standard week in the month of September. Around 76% of rainfall was confined to five months of the year (i.e., May to September). During these five months the number of rainy days exceeded more than fifteen (15). The probability distribution of amount of rainfall on standard weekly basis is shown in Figure 1.2. The average weekly evaporation is shown in Figure 1.3. The recorded data of rainfall and evaporation has a significant role in the design of the irrigation system and also for evaluating

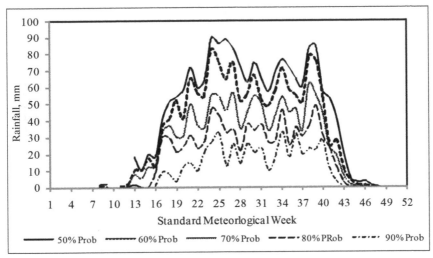

FIGURE 1.2 Probability distribution of amount of rainfall in a standard meteorological week.

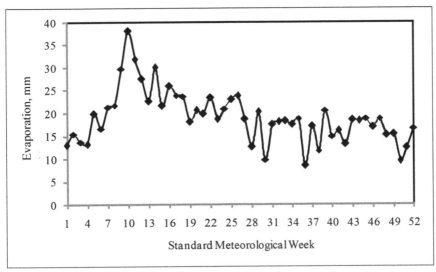

FIGURE 1.3 Weekly evaporation at the site.

the water requirement of the crops grown under any irrigation systems. The irrigation scheduling can also be done based on the assured value of these meteorological parameters.

1.2.2 SOIL ANALYSIS AND SOIL MOISTURE CHARACTERISTIC CURVE

Soil samples were collected using the standard procedure at a depth of 0–20 cm prior to growing of vegetables. The soil samples collected were air-dried, ground and pass with 2 mm sieve. For soil organic carbon analysis soil sample was sieved with 0.5 mm. The methods used to analyzes the physico-chemical properties of soil and their average recorded values are presented in the Table 1.1. The values of the soil pH was moderately acidic in reaction (pH-5.4), soil texture is sandy clay (64.3% sand, 14.7% silt and 21% clay); soil available nitrogen (211 kg ha⁻¹) was low (less than 280 kg.ha⁻¹); medium in available phosphorus (18.3 kg.ha⁻¹) and potassium (156 kg.ha⁻¹); high in organic carbon (1.89%) (more than 0.5%) and iron toxicity was present. The soil microbial biomass carbon (124.8 μg g⁻¹ soil) was found to be in a medium range. A soil moisture characteristics curve was also prepared using pressure plate apparatus. The soil moisture characteristics curve will enable to know the moisture holding capacity of the soil and accordingly the irrigation scheduling can be prepared (Figure 1.4).

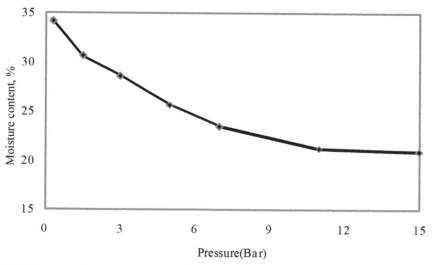

FIGURE 1.4   Soil moisture characteristics curve of the soil at the study area.

TABLE 1.1 Physico-Chemical Properties of Soil

Particulars	Methods used	Avg. Value	Inferences	Ref.
Available K_2O (kg ha^{-1})	Flame photometric determination	156	Medium	[5]
Available N (kg ha^{-1})	Alkaline Permanganate Method	211	Low	[15]
Available P_2O_5 (kg ha^{-1})	Bray's method	18.3	Medium	[5]
Organic carbon (%)	Walkley and Blacks method of rapid titration	1.11	High	[8]
pH	Soil: Water (1:2.5), pH meter with glass electrode	5.4	Moderately acidic	[5]
Soil microbial biomass carbon (μg g^{-1} soil)	Chloroform fumigation extraction method	124.8	–	[2]
Soil texture	Hydrometer method	–	Sandy clay	[4]

1.2.3 LAYOUT OF THE GRAVITY-FED DRIP IRRIGATION (GDIS) PLOT

The trail was conducted at the site with irrigation water supply from GDIS and hand watering irrigation in control plot (without GDIS) located nearby. The projected area of the site is around 1,600 square meters, where the system components of the gravity-fed drip irrigation system (GDIS) were installed. Around 500 square meter of adjacent land was also cultivated with the winter vegetables under control plot (i.e., with hand watering practices). Four winter vegetables were experimented with the GDIS and the land allocated to each vegetable is around 400 m^2. Similarly in the control plot for each crop was around 125 m^2.

The plots under GDIS were irrigated by gravity-fed drip irrigation system, whereas the control plots were irrigated using hand watering. The field layout and cultural operations are shown in Figure 1.5.

The performance of the crops, *viz.*, Cabbage, Broccoli, Cauliflower and Babycorn was also recorded both for trial plot and control plots. Nursery bed was prepared except for Baby corn at the site for Cabbage, Broccoli and Cauliflower. Around 20–25 days tender plant saplings were transplanted at the study area in the pits previously prepared and manured with

A. Layout of pipeline on the field

B. Layout of lateral on the sloppy terrain

C. Installation of a water tank on the
mid hill for operating the gravity-fed
drip irrigation system

D. Installation of a filtering unit and
other accessories

FIGURE 1.5   Field layout and cultural operations.

Farm Yard Manure (FYM). As the farmers of this NEH regions mostly go for boon methods of cultivation (A traditional practice of cultivating in the sloppy terrain for *khasi* tribal farmers of Meghalaya, where the cultivable lands were made strips raised beds with a gap of 50–60 cm between two raised bed strips, so that stagnation of water in the sloppy area is not ensured). The tribal farmers are not in the favor of terrace cultivation (except in some patches where terrace cultivation is predominant), as the preparation of terrace will cost a huge amount. The detailed crop calendar of the crops is presented in Table 1.2.

The field layout for the gravity-fed drip irrigation system and the crop performances is shown in Figure 1.6.

No modifications of the boons were done at the farmer's field, to raise the crop under GDIS. The laterals were laid along the boon (raised strip beds) with online drippers placed at suitable distance apart. The plant spacing was also maintained based on the standard package of practices.

TABLE 1.2 Detailed Crop Calendar for the Crops Grown Under GDIS

Field operation	Cabbage	Broccoli	Cauliflower	Baby corn
	Date of Operation			
Cultural Operations				
Nursery bed preparation	23rd Oct. 2014	20th Nov. 2014	3rd Nov. 2014	–
Field preparation	30th Oct. 2014	3rd Jan 2015	5th Jan 2015	22nd Oct. 2014
Field layout	30th Oct. 2014	3rd Jan 2015	5th Jan 2015	22nd Oct. 2014
Sowing/ planting	1st Nov. 2014	6th Jan. 2015	8th Jan 2015	25th Oct. 2014
FYM Application	30th Oct. 2014	3rd Jan 2015	5th Jan 2015	22nd Oct. 2014
Intercultural operations				
Gap filling	10th Nov. 2014	28th Jan 2015	–	15th Nov. 2014
Weeding	12th Dec. 2014	5th Feb 2015	5th Feb 2015	20th Nov. 2014
	15th Jan. 2015	12th March 2015	12th March 2015	22nd Dec. 2014
	5th Feb. 2015			
Crop protection measure	15th Jan. 2015	12th Feb 2015	12th Feb 2015	–
Date of harvesting	13th March 2015	2nd April 2015	17th March 2015	5th March 2015

Golden acar certified seed variety was used for Cabbage at the rate of 0.6 kg per hectare at a planting geometry of 60 × 45 cm^2; *King of market* certified seed variety was used for Broccoli at the rate of 0.75 kg per hectare at a planting geometry of 60 × 45 cm^2; *Puspa dipali* certified seed variety was used for Cauliflower at the rate of 0.5 kg per hectare at a planting geometry of 60 × 45 cm^2; similarly a local variety of Baby corn was used at the rate of 12.5 kg per hectare at a planting geometry of 60 × 60 cm^2.

As the farmers of these NEH regions are not in favor of raising the crops with chemical fertilizers, a standard blanket application of farm yard manure (FYM) at the rate of 15 tons per hectare was applied as basal for all the winter crops under investigation. Gap filling and weeding operations were done as and when required as a part of intercultural operations

A. Planting of seeding in pits

B. Cabbage crop at the site

C. On-farm Farmers interaction
with standing crop Babycorn

D. Beneficiary at the GDIS plot

FIGURE 1.6 Field layout for the gravity-fed drip irrigation system and the crop performance.

of the winter vegetables. The details dates of intercultural operations are presented in Table 1.2. As a part of crop protection method, neem-based pesticides were used for necessary spraying both in the trial and control plots. Harvesting of the crops was done in the 2nd week of March and 1st week of April for Broccoli.

1.2.4 IRRIGATION SCHEDULING AND MEASUREMENT OF STREAM DISCHARGE

Discharge of the perennial stream feeding the water tank of the GDIS system installed at the site was recorded during pre-monsoon, monsoon and post monsoon season. For this exercise, the materials required include a hose pipe, 20 liter bucket, digital stopwatch and a note book for noting the data. The hose pipe was formerly fitted at a particular location of the stream at higher elevation. The other end opens up for necessary collection of water at the storage tank. For measuring the volume of water discharged from the stream, the hose pipe was put in a bucket of sufficient capacity. The volumetric measurement of the stream water discharge during the experimental period is shown in Figure 1.7. Simultaneously, the stopwatch was started to accurately maintain the time limit (1 minute). This step was repeated for three consecutive observations and average of three reading was taken as the discharge of the stream during the particular season. The average recorded discharge was found from the perennial stream varied between 16.59 to 17.95 liters per minute. The maximum recorded discharge was 20.55 liters per minute during monsoon period. The average stream discharge was presented in Table 1.3. The water from the perennial stream has been tapped using a suitable plastic conveyance pipe and the tapped water was allowed to fill a poly vinyl chloride (PVC) tank of 1000-liters capacity. The number of fillings of the PVC tank during the cropping season was also recorded. Irrigation was scheduled at moisture depletion pattern of the soil. Irrigation was provided by the GDIS when the soil moisture content fell below 50% of the available moisture holding capacity of the soil. Hand watering was practiced in the control plot using a garden watering can of 10-liters capacity. The volume of water applied was recorded and finally the water used by the crops from the GDIS plots and control plots was arrived at.

A. Volumetric measurement of spring water discharge

B. Volumetric measurement of spring water discharge before feeding to the water tank

FIGURE 1.7   Volumetric measurement of the stream water discharge.

TABLE 1.3   Average Stream Discharge Recorded at the Site

Stage	Average discharge volume (L min⁻¹)
Pre-monsoon	16.59
Monsoon	17.95
Post-monsoon	17.56

1.2.5 UNIFORMITY COEFFICIENT OF THE GDIS

For evaluating the performance of the GDIS, the discharge from the drippers was recorded to calculate the uniformity of application. Online dripper discharge along the laterals was also recorded to study the variation in discharge of water distributed along the slope of the study area. Similarly, the dripper discharge across the laterals was also recorded to study the variation of water gradually away from the tank. For evaluating the online dripper discharge, authors used plastic containers of 500 ml capacity, measuring cylinders (250 ml and 500 ml) and a digital stopwatch. The plastic containers were placed directly below each dripper along the lateral. The valve of the mainline of the GDIS was turned on. Simultaneously, the stopwatch was started to record the time. The valve was turned off after one minute. Water collected in the containers was then transferred to the measuring cylinder for recording the total volume from each dripper. This process was repeated for three times and an average value of discharge from the respective dripper was evaluated.

Similarly to record the discharge of dripper across the laterals, the containers were placed directly below each dripper of the lateral at the same level across the slope and the volume of water per minute was recorded. The volumetric discharge measurements from the dripper are shown in Figure 1.8. The uniformity coefficient of the GDIS was determined using standard protocol given in Eq. (1).

$$Cu = 100 \, [1.0 - \Sigma \, X \, / \, (m \times n)]) \tag{1}$$

A. Dripper discharge measurement for the gravity-fed drip irrigation system

B. Volumetric measurement of dripper discharge using a measuring cylinder

FIGURE 1.8 Volumetric discharge measurement from the dripper.

where, Cu = uniformity coefficient, percentage; m = average value of all application depths, mm; n = total number of observation points; and X = average numerical deviation of individual observations from average application rate in mm.

1.3   RESULTS AND DISCUSSION

The crop biometrics and yield attributes were recorded from the tagged sample plants and accordingly the yield data were converted to yield per hectare. The productivity of the winter vegetables as recorded form the representative cultivated land of 400 m² area may not represent the project yield (t ha⁻¹). As the gravity-fed drip irrigation system has its limitations of supplementing irrigation needs to small patches of land generally < 1 to 1.5 acre [3, 6]. The crop performance and the system performance in terms of uniform distribution of irrigation water in the cultivated land are described in the subsequent sections.

1.3.1   CROP PERFORMANCE UNDER GDIS

Crops taken up for the trial included Baby corn, Cabbage, Broccoli and Cauliflower in the farmer's field. The trial was also conducted with a control plot (where hand watering was provided for irrigation). The irrigation scheduling was done based on 50% moisture depletion. These winter crops are high value crops and generally fetch a good market since these are grown under organic inputs. Table 1.4 indicates that the yield recorded

TABLE 1.4   Yield Recorded for the Winter Vegetables under GDIS and Control Plots

Crop	GDIS trial plot	Control plot
	Tons/ha	
Baby Corn	2.8	2.0
Broccoli	11.8	9.6
Cabbage	12.6	10.4
Cauliflower	12.4	8.6

from Babycorn, Cabbage, Broccoli and Cauliflower was 2.8, 12.6, 11.8 and 12.4 tons-ha⁻¹ for the GDIS plots and 2.0, 10.4, 9.6 and 8.6 tons-ha⁻¹ for control plots. It may be observed that winter vegetables cultivated using GDIS not only increased the income and but also added extra food to the food basket, and it also increased the cropping intensity of the state when taken up in a large scale.

The performance of cabbage and cauliflower are better compared to baby corn. All types of Cole crops can be taken up under this GDIS in the undulating topography of NEH region. The trials could not include liming application to ameliorate the acidic condition of the cultivated land due to disagreement of the beneficiary farmers. However, under non-acidic conditions, the performance of the winter crop could have been better when taken up with inorganic fertilizer inputs and other package of practices.

The depth of water required for growing the crops during its crop period was recorded on volumetric basis for GDIS and control plot and is presented in Table 1.5. It may be noted that for all crops the amount of water use was more under control plot as compared to the GDIS plot. Under the circumstances of evaluating the effectiveness of the GDIS technology, the judicious water use may be ascertained by evaluating the water use efficiency (WUE), which is the ratio of recorded yield to the quantum of water use. The WUE was also calculated for the respective crop (Table 1.5). It may be noted from the Table 1.5 that the WUE was found to be the maximum for cauliflower and minimum for baby corn. The WUE was found better for the GDIS trial plot compared to the control plot.

TABLE 1.5 Crop Water Use Efficiency for Winter Vegetables

Crop	Plot	Delta (cm)	WUE (kg ha⁻¹ cm⁻¹)
Baby corn	GDIS trial	455	6.15
	Control	495	4.04
Cabbage	GDIS trial	560	22.50
	Control	625	16.64
Broccoli	GDIS trial	662	17.82
	Control	685	14.01
Cauliflower	GDIS trial	685	18.10
	Control	715	12.03

1.3.2   UNIFORMITY COEFFICIENT UNDER GDIS

The dripper discharges were recorded for the dripper numbered from 1 to 15 along the slope. The variation of the discharge as recorded is shown in Figure 1.9. The discharge variation in the drippers is more than 20% when

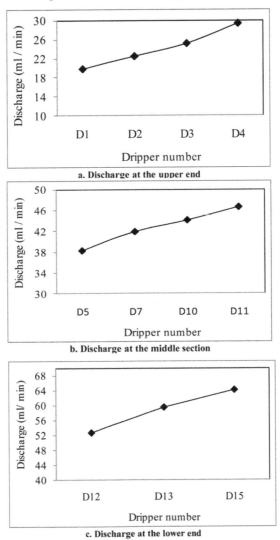

a. Discharge at the upper end

b. Discharge at the middle section

c. Discharge at the lower end

FIGURE 1.9   Dripper discharge along the lateral as recorded at three sections on the lateral: (a) Discharge at the upper end; (b) Discharge at the middle section; (c) Discharge at the lower end.

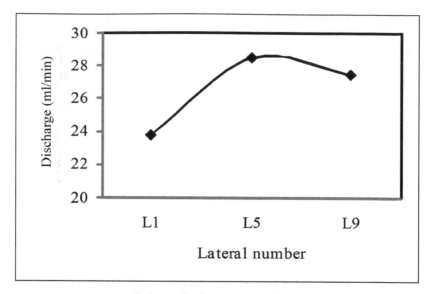

a. Dripper discharge near to the tank

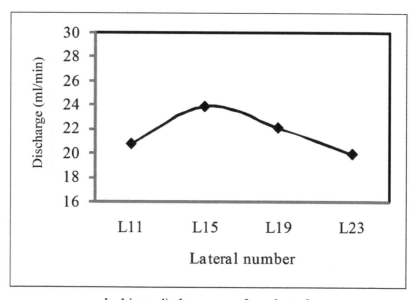

b. dripper discharge away from the tank

FIGURE 1.10 Dripper discharge across the laterals away from the irrigation water supplying tank: (a) Dripper discharge near to the tank; (b) Dripper discharge away from the tank.

traveled from upstream of the sloppy land to the lower end. The discharge was increased along the slope steadily due to availability of virtual pressure head at the lower end of the sloppy land.

The increase of discharge can be compensated in the undulating topography by using pressure-compensating drippers. However, using the lateral lines across the slope may nullify the variation to a greater extent.

Similarly for the same dripper, the discharge was recorded across the lateral lines in the site. The variation of dripper discharge across the laterals away from the irrigation water supplying tank is shown in Figure 1.10. It may be noted that across the lateral for the same dripper also there is an increase in discharge, this may be due to a natural gradient exist in the trial plot away from the irrigation water tank to a certain extent and after that the grade was nullified. The dripper discharge was found more away from the tank to a certain point up to lateral #9 and thereafter the dripper discharge again decreases towards the far end.

Due to larger variation of the discharge recorded in the sloppy cultivated land, the uniformity coefficient of the system was found to vary between 64 and 88%.

It was found that the variation of uniformity coefficient varied between 64–88%. The discharge was found to be more down the slope than at the upper side. Hence, to achieve a better uniformity higher discharge dripper may be put on the lateral at the upper end of the sloppy land and lower discharge dripper may be installed at the lower end of the lateral to ensure uniformity of the discharge. Pressure compensating type of drippers may be used to nullify the discharge variation to a greater extent.

1.4 CONCLUSIONS

Irrigation water management in the undulating hilly topography is always a challenging job, where judicious water usage is not ensured. Irrigation being a lifeline of agriculture suitable low cost, semi- mechanized irrigation technology like gravity-fed drip irrigation system has a greater role to play in irrigation management of hilly areas. This technology is socially acceptable and financially feasible option. This irrigation technology enhances the productivity and thereby cropping intensity. The gravity fed drip irrigation system (GDIS) is a low cost irrigation system, mostly suited for the hilly terrain, where the irrigation system can be operated without any external power requirement. It is suitably for small areas (mostly 1.5 acres). There is every possibility of raising winter crops during non-rainy seasons which will increases the farm income and thereby increases the cropping intensity and production.

1.5   SUMMARY

To feed the increasing global population, water sector has a key role to play. As agriculture consumes a lion's share of "water," its non-judicious usages leads to multi facials problems. Micro irrigation systems are the most efficient methods, but the adherence of these technologies is under threat due to some reasons or others. Gravity-fed drip irrigation system (GDIS) is one of the improvisations of these technologies where the pumping cost is nullified and can be brought under practice in undulating topography. It is a low cost irrigation technology operated by the force of gravity and applicable for small patches of arable land in hilly regions. Water for operating the system can be tapped either from perennial springs or harvested at the mid hills using some lining material in a small reservoir. Performance of winter vegetables was studied under GDIS to assess the feasibility of gravity-fed drip irrigation system with baby corn, cabbage, broccoli and cauliflower as trial crops in hilly terrain of Meghalaya during 2014–2015 *rabi* season. It was recorded that the yield of baby corn, cabbage, broccoli and cauliflower was 2.8, 12.6, 11.8 and 12.4 tons-ha^{-1} for the GDIS plots and 2.0, 10.4, 9.6 and 8.6 tons-ha^{-1} for control plots, in the trial areas. The uniformity coefficient of the system varied from 64 to 88%.

ACKNOWLEDGEMENT

The financial assistance from Department of Science and Technology, Water Technology Initiatives (DST, WTI) – Government of India, New Delhi (DST/WTI/2K12/37(G)) for conducting the experiment is thankfully acknowledged by the authors.

KEYWORDS

- drip irrigation
- dripper
- gravity-fed drip irrigation system
- India
- low head drip irrigation
- micro irrigation
- North Eastern Hill region
- North Eastern India
- on line dripper
- spring water tapping
- undulating topography
- uniformity coefficient
- volumetric water measurement
- water use efficiency
- winter vegetables

REFERENCES

1. Bhatnagar, P. R., & Srivastava, R. C. (2003). Gravity-fed drip irrigation system for hilly terraces of the northwest Himalayas. *Irrig. Sci.*, *21*, 151–157.
2. Brookes, P. C., & Joergensen, R. G. (2006). Microbial biomass measurements by fumigation extractant. In: Bloem, J., Hopkins, D. W., & Benedetti, A. (eds.). *Microbiological Methods for Assessing Soil Quality*. CABI Publishing, Oxfordshire, UK, pp. 77–83.
3. Chigerwe, J., Manjengwa, N., van der Zaag, P., Zhakata, W., & Rockstrom, J. (2004). Low head drip irrigation kits and treadle pumps for smallholder farmers in Zimbabwe: a technical evaluation based on laboratory test. *Phy. Chem. Earth*, *29*, 1049–1059.
4. Gee, G. W., & Bauder, J. W. (1986). Particle size analysis. In: Klute, A. (ed.). *Methods of Soil Analysis. Part 1. 2nd Edition*. Agronomy Monograph 9, ASA and SSSA, Madison, WI, pp. 383–411.
5. Jackson, M. L. (1973). *Soil Chemical Analysis*. Prentice Hall of India Pvt. Ltd., New Delhi, India, pp. 485.

6. Karlberg, L., Rockstrom, J., Annandale, J. G., & Steyn, J. M. (2007). Low-cost drip irrigation: A suitable technology for southern Africa? An example with tomatoes using saline irrigation water. *Agri. Water Man.*, 89, 59–70.

7. Senzanje, N. A., Rockstrom, J., & Twomlow, S. J. (2005). On farm evaluation of the effect of low cost drip irrigation on water and crop productivity compared to conventional surface irrigation system. *Phy. Chem. Earth* (Parts A/B/C), *30*(11–16), 783–791.

8. Piper, C. S. (1966). *Soil and plant analysis*. Hans Publisher, Bombay, India.

9. Rajput, T. B. S., & Patel, N. (2001). Moisture front advance studies under drip irrigation. *Paper Presented in 88th Indian Science Congress*, Jan. 3–7, held at IARI, New Delhi.

10. Ray, L. I. P., Clarence, G. K., Singh, N. J., Singh, A. K., & Bora, P. K. (2014). Gravity-fed Drip Irrigation for Hilly Region. *Published by Dean, College of Post Graduate Studies (CAU)*, Umroi Road, Umiam – 793103, Meghalaya.

11. Rawlins, S. L. (1977). Uniform irrigation with a low-head bubbler system. *Agri. Water Man.*, 2(1), 167–178.

12. Reynolds, C., Yitayew, M., & Petersen, M. (1995). Low-head bubbler irrigation systems. Part I: Design. *Agri. Water Man.*, 29, 1–24.

13. Reynolds, C., & Yitayew, M. (1995). Low-head bubbler irrigation systems. Part II: Air lock problems. *Agri. Water Man.*, 29, 25–35.

14. Singh, S. D., and Singh, P. (1978). Value of drip irrigation compared with conventional irrigation for vegetable production in a hot and arid climate. *Agron. J.*, *70*(6), 945–947.

15. Subbiah, B. V., & Asija, G. L. (1956) Rapid procedure for estimation of available nitrogen in soil. *Current Science*, *25*, 259–260.

16. Yao, W. W., Ma, X. Y., Li, J., & Parkes, M. (2011). Simulation of point source wetting pattern of subsurface drip irrigation. *Irrig. Sci.*, *29*, 331–339.

17. Yazar, A., Sezen, S. M., & Sesvern, S. (2002). LEPA and trickle irrigation of cotton in the Southeast Anatolia Project (GAP) area in Turkey. *Agri. Water Man.*, *54*, 189–203.

18. Yitayew, M., Didan, K., & Reynolds, C. (1999). Microcomputer based low-head gravity-flow bubbler irrigation system design. *Computers and Electronics in Agriculture*, *22*, 29–39.

CHAPTER 2

PERFORMANCE OF DRIP IRRIGATED CAPSICUM UNDER PROTECTED CULTIVATION STRUCTURES

L. N. SETHI and H. CHAKRABORTY

CONTENTS

2.1 INTRODUCTION

The North-Eastern region (NER) of India is a land of magnificent beauty, possessing undulating hills, rolling grasslands, cascading waterfalls, snaking rivers, terraced slopes and thrilling flora and fauna. This picturesque scenario is contrasted by widespread poverty, low per capita income, high unemployment and low agricultural productivity leading to food-insecurity. The high vulnerability to natural calamities like floods, submergence, landslides, soil erosion, etc., has resulted in low and uncertain agricultural productivity. The low utilization of modern inputs in agriculture has further reduced the ability of the farm households to cope with high risks in production and income [6]. Therefore, there is urgent need for adoption of advance technologies such as protective cultivation and suitable irrigation methods to protect soil fertility and increase the agricultural productivity.

The 20th century brought significant changes to the economics of global agriculture. This transition is a direct result of the increase in relative price of labor and changes in domestic and global agricultural policies [3, 19], and was spurred by dramatic improvements in agricultural productivity, and a shift from more labor intensive agriculture to more capital- and technology intensive agricultural practices that employed new varieties, synthetic inputs, and irrigation [2, 7–11, 18, 20]. Incorporating and disseminating technological advances that improve productivity and incomes in smallholder farming systems, remains a challenge throughout the developing world [4]. India being a vast country with diverse and extreme agro-climatic conditions, the protected vegetables cultivation technology can be utilized for year round and off-season production of high value, low volume vegetables, crops production of virus free quality seedlings, quality hybrid seed production and as a tool for disease resistance breeding programs.

Vegetables are generally sensitive to environmental extremes, and thus high temperatures and limited soil moisture are the major causes of low yields and will be further magnified by climate change. India is the second largest producer of vegetables in the world, next to China. India's share of the world vegetable market is around 14%. It produces 133.5 million tons of vegetables from an area of 7.9 million hectares. According to statistics release by Ministry of Agriculture, there has been 13.5% increase in area and 13.4% increase in vegetable output during the period 1996 to 2010 that indicates the need of research to increase the vegetable production against the available area with advanced technology of cultivation.

Capsicum (*Capsicum annuum* L.) is also called as bell pepper or sweet pepper and is one of the most popular and highly remunerative annual herbaceous vegetable crops. Capsicum is cultivated in most parts of the world, especially in temperate regions of Central and South America and European countries, tropical and subtropical regions of Asian continent mainly in India and China. India contributes one fourth of world production of capsicum with an average annual production of 0.9 million tons from an area of 0.885 million hectare with a productivity of 1266 kg per hectare. In India, capsicum is extensively cultivated in Andhra Pradesh, Karnataka, Maharashtra, Tamil Nadu, Himachal Pradesh, and hilly areas of Uttar Pradesh. Andhra Pradesh stands first in area of 236,500 ha with a production of 74,850 tons. And Karnataka stands second in area of about 76,000 ha with a production of about 131,000 tons [1].

Enhancing and sustaining the productivity on hillocks in northeastern region of India is a major challenge as it is practiced under ecologically fragile environments which include altitudinal, climatic and topographical variations. In spite of the great importance of vegetable crops, it faces a lot of constraints like photo stress, moisture stress, temperature stress, and weeds growth, deficiencies in soil nutrients, excessive wind velocities and atmospheric carbon-dioxide. These constraints could be alleviated by adopting a unique, specialized hi-technology for protected cultivation with efficient irrigation method.

The trend in recent years has been towards conversion of surface irrigation to drip irrigation to improve plant quality and yield. While, in present, some farmers are not sure when and how much water they should irrigate under drip irrigation condition, and they tend to confirm irrigation timing and amount according to conventional experience, and then, induce new water loss under new technology. Drip irrigation can distribute water uniformly, precisely control water amount, increase plant yields, reduce evapotranspiration (ET) and deep percolation, and decrease dangers of soil degradation and salinity [3, 5, 12]. Therefore, an easy-operation irrigation scheduling method is very stringent for capsicum with drip irrigation condition.

Protected cultivation is a cropping technique wherein the micro environment surrounding the plant body could be controlled partially/fully as per plant need during their period of growth to maximize the yield and resource saving. Greenhouse is one of the most practical methods of achieving the objectives of protected agriculture, where natural environment is modified by the use of sound engineering principles to achieve optimum plant growth and yield (more produce per unit area) with increased input use efficiency [16, 17]. The protected cultivation of vegetable offers distinct advantage of quality, productivity and favorable market price to the growers. It increases their income in off- season as compared to normal season. Off season cultivation is one of the most profitable technologies. Virus free cultivation of Tomato, Chili, Capsicum, cucumber and other vegetables are essential chiefly during rainy season.

Therefore in the present study, a field experiment was carried out with high cost greenhouse and low cost shade net protective cultivation structure for capsicum production with drip irrigation system to control the environment by providing protection from the excessive heat, rain and cold and also increase the crop productivity.

2.2 MATERIALS AND METHODS

2.2.1 EXPERIMENTAL SITE

The experiment was conducted at experimental field located in hilly terrain of Department of Agricultural Engineering, Assam University, Silchar, Assam, India, during October 2014 to April 2015. The experimental field is situated at 24°41′ N latitude and 92°45′ E longitude at an elevation of 41 meters from the mean sea level. Figure 2.1 shows the location of the experimental site.

The experiment was laid out in three different blocks: greenhouse, shade-net house and open field. A field experiment was carried out with protected cultivation structures such as Hi-tech poly greenhouse, low-cost bamboo structure shade-net house and an open field two types of vegetable crops (Capsicum). Each plot with capsicum crop for an area of 20 (4×5) m² was selected for experimentation. Each plot had eight rows and four replications.

The climate of the north eastern region is subtropical, warm and humid. The average rainfall of the region is 3180 mm with average rainy days of 146 days per annum. In each block, soil physical, chemical and nutrient analysis of soil was carried to find out the initial status and irrigation, nutrient requirement for crop growth in the experimental plots.

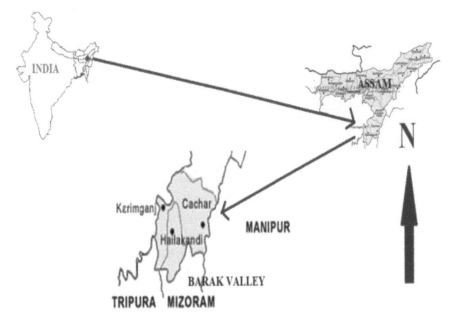

FIGURE 2.1. Location of experimental site.

2.2.2 *PROTECTED CULTIVATION STRUCTURES*

Protected cultivation practices is a cropping technique wherein the micro environment surrounding the plant body is controlled partially or fully as per plant need during their period of growth to maximize the yield and resource saving. Greenhouse, Poly house, Poly tunnel and Net house protect the crop from high intensity of light, high rainfall, winds, insects through structure, polyethylene film or polycarbonate sheet, shading nets, insect net, cooling pad, exhaust fan, foggers and drip irrigation systems, fertigation equipment etc., which controls light, temperature, humidity and irrigation with fertigation and other required growth substances directly into the root zone of the plant. However, high-tech greenhouse and low cost bamboo structure shade-net house were selected for the cultivation of capsicum for the present study based on suitability, local conditions and availability of resources, cost effectiveness and requirement.

2.2.2.1   Greenhouse

Greenhouses are climate controlled and have a variety of applications, the majority being, off-season growing of vegetables, floriculture, planting material acclimatization, fruit crop growing for export market and plant breeding and varietals improvement. Greenhouses are built of a G.I. structure wherein crops are grown under a favorable artificially controlled environment and other conditions viz. temperature, humidity, light intensity, photo period, ventilation, soil media, disease control, irrigation, fertigation and other agronomical practices throughout the season irrespective of the natural conditions outside. The greenhouse is generally covered by transparent or translucent material such as glass or plastic. The greenhouse covered with simple plastic sheet is termed as poly house. The greenhouse generally reflects back about 43% of the net solar radiation incident upon it allowing the transmittance of the "photo synthetically active solar radiation" in the range of 400–700 Nm wave length.

FIGURE 2.2   View of a greenhouse structure selected for experiment in the study site.

Figure 2.2 shows the view of a greenhouse selected for the field experiment in the study site. It is a high-tech greenhouse of tubular structure covered with 200 micron UV film and shade net, which is designed to withstand wind up to 120 km/hr., and trellising loads up to 25 kg/m2, with 4-way fogger irrigation system and cooling system inside the greenhouse by foggers, cooling pads and exhaust fans.

2.2.2.2 Shade-Net House

A net house are basically naturally ventilated climate controlled and has a variety of applications, the majority being, growing of vegetables, floriculture, and fruit crop growing for export market. Figure 2.3 shows the view of a shade-net house installed in the study site. The shade-net house is a low cost bamboo structured house of height 2.2 m with 50% shading shade nets with all the sides and ceiling covered with green shade net. The shade net of Green x Green – enhance the process of photosynthesis in plants resulting better foliage in ornamental plants.

FIGURE 2.3   View of a shade-net house structure constructed in the study site.

2.2.3 DRIP IRRIGATION SYSTEM FOR CAPSICUM

Water management and water scarcity in winter season is one of the major constraints on vegetable production in hilly terrain of North East. Soil moisture is one of the predominant factors influencing tomato and capsicum productivity and drip irrigation is the best alternative. The water use efficiency (WUE) of drip irrigation is 90–95%, whereas sprinkler has 70–80% and surface irrigation has 30–50% WUE.

Drip irrigation involves technology for irrigating plants at the root zone through emitters fitted on a network of pipes (mains, sub-mains and laterals). Emitting devices could be drippers, micro sprinklers, mini sprinklers, micro-jets, misters, fan jets, micro sprayers, foggers and emitting pipes, which are designed to discharge water at prescribed rates. Water requirement, age of plant, spacing, soil type, water quality and availability are some of the factors which would decide the choice of the emitting system. Drip irrigation system can distribute water uniformly, precisely control water amount, increase plant yields, reduce evapotranspiration and deep percolation, and decrease dangers of soil degradation and salinity.

Capsicums are particularly sensitive to cold and growth is inhibited below 10°C. Rain and high humidity can increase development and spread of diseases, particularly bacterial spot. Low humidity favors mites and powdery mildew. Wind can cause fruit rub and blemish, and increase water stress, resulting in the development of the fruit disorder blossom-end rot. Capsicums grow best in deep, well-drained, medium-textured soil, such as loams, but can be grown in a wide range of soil types. Soil should be at least 30 cm deep and have good drainage. Optimum pH of 5.5–6.5 is suitable for capsicum cultivation. Hence, capsicum seed of hybrid F1 Juliet was selected with crop spacing of 0.45×0.45 m2.

Figure 2.4 shows the layout and installation of drip irrigation system at the greenhouse, shade net house and open field of the study site. In the present study a poly tank of 1000 liter capacity fed from the groundwater source, online drip irrigation system with 4 liter/hour drippers, one dripper per plant, 12 mm on-line laterals made of low density poly ethylene material and 50 mm polyvinyl chloride pipe with control valve and flush valve were selected and installed for the field experiment.

FIGURE 2.4 View of drip irrigation system installed at (a) Greenhouse, (b) Shade net and (c) Open fields for capsicum crop.

2.2.4   FIELD EXPERIMENTATION

2.2.4.1   Status of Soil Properties and Nutrient Application

In order to carry out the field experiment for crop production, the soil physico-chemical characteristics such as soil texture, bulk density, moisture content, soil pH, electrical conductivity were determined using standard method. The soil texture and physical characteristics play the important role for sustainable crop planning in an area. The soil texture is responsible for the retention of many of the plant nutrients in the soil, such as calcium, magnesium, potassium, trace elements and some of the phosphorus. The status of the soil texture and physical characteristics for three different treatment of cultivation are given in Tables 2.1 and 2.2, respectively.

The soil type observed in open field, shade net and greenhouse are sandy loam, loamy sand and sandy loam, respectively that indicate the almost similar type of soil. The clay content was higher in greenhouse as compared to other that indicates more ability to retain plant nutrients, or to release them to the soil solution for plant uptake. However, the sand content was found higher in shade net house which implies little or no ability to supply grass with nutrients or to retain them against leaching. The physical characteristics of soil status revealed that moisture holding capacity is higher in greenhouse followed by shade net and then open field, which implies the water and nutrient requirement for crop production.

Capsicums grow best in deep, well-drained, medium-textured soil, such as loams, but can also be grown in a wide range of soil types. In

TABLE 2.1   Soil Texture Status Observed in Three Experimental Blocks at the Study Site

| Experimental blocks | Soil texture | | | |
	Percentage of sand	Percentage of silt	Percentage of clay	Soil type
Greenhouse	56.1	34.8	9.1	Sandy Loam
Shade-net House	70.2	27.2	2.6	Loamy Sand
Open Field	61.4	35.9	2.7	Sandy Loam

TABLE 2.2   Soil Physical Characteristics Observed in Three Experimental Blocks at the Study Site

| Experimental blocks | Bulk density) | Moisture content | Saturated moisture content | Field capacity | Wilting point |
	g/cm^3	%			
Greenhouse	1.46	24.48	34.11	26.95	19.72
Open Field	1.77	9.14	25.86	17.23	12.54
Shade-net House	1.82	8.82	20.76	18.16	13.65

the present field experiments, the capsicum seed of hybrid F1 Juliet was selected and planted at crop spacing of 0.45 × 0.45 m². Nutrient requirement of farm yard manure at 10 tons, N 120 kg, P_2O_5 60 kg and K_2O 60 kg/ha are recommended by Assam Agricultural University for capsicum, half of N and full doses of FYM, P_2O_5 and K_2O were applied as basal and the remaining half of N to be top dressed at 30–35 days after transplanting.

2.2.4.2   Growth and Agricultural Production Economics Analysis

The classical approach in plant growth analysis was used to find out the relative growth rate with plant weight at different harvests time. The other parameters of agricultural production that are evaluated are number of branches, leaves, flowers, fruits at the time of flowering and harvesting, fruit length and width, fruit weight, yield per plant, plot and hectare.

Agricultural production economic analysis includes how economically and efficiently the production can be done. It included the cost for land preparation, nursery and seedlings preparation, manures and fertilizers, plant protection measures, labor cost, land revenue, etc., The economics of production of capsicum cultivation included the gross returns, net returns, net profit, benefit cost ratio and payback period.

2.3   RESULTS AND DISCUSSION

2.3.1   GROWTH AND PRODUCTION OF CAPSICUM WITH DRIP IRRIGATION

The field experiments for drip irrigated capsicum under two protected cultivation structures and open field on a hilly terrain were carried out from October 2014 to April 2015. The crop growth and yield parameters were monitored and analyzed.

The variations of plant height at different stages of plant growth of capsicum crop in three experimental blocks of cultivation are presented in Figure 2.5. It was observed that the greenhouse produced the plants with higher plant height on an average but during few growth stages the plant height goes higher in shade-net house.

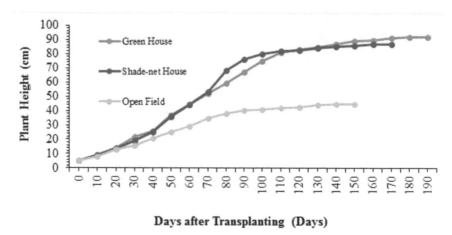

Days after Transplanting (Days)

FIGURE 2.5   Variation of plant height at different growth stage of capsicum in three conditions of cultivation.

The greenhouse cultivation produced the tallest plant (92 cm) on an average and the shortest plant (45 cm) was obtained from the cultivation in open field at final harvesting stage which was 51% higher. The results of the present study are in agreement with the findings of Maya et al. [15] who stated that, plant height of sweet pepper was significantly increased with close spacing. Manchanda et al. [14] also expressed similar observations on plant height of capsicum. Also the duration of crop was maximum in greenhouse cultivation (44 days more) followed by shade-net cultivation (23 days more) than the open field cultivation.

The maximum average number of branches per plant, number of leaves per plant, 1st day of flowering and 1st and last day of harvesting were found almost nearer in case of greenhouse and shade-net house cultivation but it differed significantly in open field cultivation (Table 2.3).

Maximum average number of branches (11 per plant) was recorded under greenhouse cultivation and the lowest number of branches (8 per plant) was recorded in the open field cultivation. The day to 1st flowering occurred at 3 days and 7 days early (47 days) in greenhouse cultivation, than shade-net house (50 days) and open field (54 days) cultivation, respectively, but the day to 1st harvesting is done 4 days and 16 days early (93 days) in open field cultivation than shade-net house (97 days) and greenhouse cultivation (109 days), respectively.

TABLE 2.3 Vegetative Growth Parameters for Capsicum in Three Conditions of Cultivation

Experimental blocks	No. of branches per plant	No. of leaves per plant	Occurrence of events (days after transplanting)		
			1st flowering	1st harvesting	Last harvesting
Greenhouse	11	45	47	109	192
Shade-net House	10	42	50	97	171
Open Field	8	34	54	93	148

It indicates that the vegetative growth was more vigorous in greenhouse and shade-net house cultivation as compared to open field cultivation and due to the hot and dry climate, the plants grown in open field matured faster and hence resulted the short crop duration. For the same reason in case of open field condition of cultivation, the temperature and solar radiation values were higher and affected directly the plant body and hence it resulted in early maturity of crop with shorter plant height, lesser number of branches and leaves but with larger stem diameter. Greenhouse cultivation resulted in highest number of fruits in this study followed by shade-net house and open field cultivation as shown in Table 2.4.

Increase in fruit number is the most important factor in yield increase. Moreover, a uniform supply of soil water throughout the growing season with drip irrigation system is needed to prevent poor fruit size and shape and to increase yield. Highest mean fruit weight was obtained from greenhouse cultivation. As the growth in all respects like number of branches,

TABLE 2.4 Average Number of Fruits Per Plant of Capsicum Per Plot Observed in Three Conditions of Cultivation

Conditions of cultivation	No. of fruits per plant in each replication				Average no. of fruits per plant
	R1	R2	R3	R4	
Greenhouse	14.33	13.97	13.82	14.01	14.03
Shade-net House	12.50	12.24	11.56	11.84	12.04
Open Field	7.50	7.67	8.00	8.17	7.83

number of flowers and plant height was higher in greenhouse cultivation, therefore the number of fruits per plant and the plant age was maximum due to the controlled atmospheric conditions with pest and weed controlled and supply of irrigation water through drip irrigation system. Number of fruits was 1.8 times higher inside greenhouse and 1.5 times higher under shade-net house than the open field cultivation.

The variations in fruit size of capsicum due to the effects of different treatments are presented in Table 2.5. The highest mean fruit length and width was obtained in greenhouse cultivation which affected the overall fruit size and weight of individual fruit.

A quite longer and wider fruits (86.21 and 68.42 mm) were obtained in the greenhouse cultivation. The open field cultivation produced the shortest and narrowest fruits (66.73 mm and 60.69 mm) and the shade-net house produced the medium length and width fruits (77.64 mm and 64.19 mm), which resulted 31% and 19% larger fruit size under greenhouse and shade-house cultivation of capsicum than the open field cultivation, respectively.

The weight of individual fruits in different types of cultivation is presented in Table 2.6. It revealed that the average weight of the individual fruit was found more (28%) in greenhouse and shade-net house (23%)

TABLE 2.5 Average Fruit Length and Width of Capsicum Observed in Three Conditions of Cultivation

Conditions of cultivation	Length of fruit (mm)	Width of fruit (mm)
Greenhouse	86.21	68.42
Shade-net House	77.64	64.19
Open Field	66.73	60.69

TABLE 2.6 Average Fruit Weight of Capsicum Observed in Three Conditions of Cultivation

Conditions of cultivation	Weight of fruit (g) in each replication				Average weight of fruit (g)
	R1	R2	R3	R4	
Greenhouse	89.43	95.61	92.19	94.73	92.99
Shade-net House	91.89	86.91	85.83	88.46	88.27
Open Field	67.35	70.77	66.67	65.84	67.66

than the open field, may be due to the controlled environmental effect which the plants received inside the protected structures of capsicum crop. Yield per plant in average was significantly influenced by protected cultivation structures than the open field cultivation (Table 2.7). The maximum yield per plant was recorded in the greenhouse cultivation which differed slightly from the shade-net house cultivation but significantly differed in case of open field cultivation.

The results revealed that approximately 2.5 times more yield per plant was observed from greenhouse and 2 times more yield from shade-net house than the open field cultivation, thus more yield can be obtained by adopting protected cultivation structures.

Protected cultivation had greater effect on yield per plot and quintal per hectare (Table 2.8 and Figure 2.6, 100 kg = quintal). The maximum yield of fruit in kg per plot of 10.44 kg/plot may led to production of 521.84 q/ha for the greenhouse cultivation than the open field yield of 4.24 kg/plot, i.e., 211.87 q/ha. It was observed that almost 146 and 101% more yield of capsicum could be obtained from the greenhouse and shade-net

TABLE 2.7 Average Fruit Yield of Capsicum Per Plant Observed in Three Conditions of Cultivation

Conditions of cultivation	Weight of fruit per plant (kg/plant) in each replication				Average weight of fruit/plant (kg)
	R 1	R 2	R 3	R 4	
Greenhouse	1.28	1.34	1.27	1.33	1.30
Shade-net House	1.15	1.06	0.99	1.05	1.06
Open Field	0.51	0.54	0.53	0.54	0.53

TABLE 2.8 Average Fruit Yield of Capsicum Per Plot Observed in Three Conditions of Cultivation

Conditions of cultivation	Weight of fruit per plot (kg/plot) in each replications				Average weight of fruit/plot (kg/plot)
	R1	R2	R3	R4	
Greenhouse	10.25	10.69	10.19	10.62	10.44
Shade-net House	9.19	8.51	7.94	8.38	8.50
Open Field	4.04	4.34	4.27	4.30	4.24

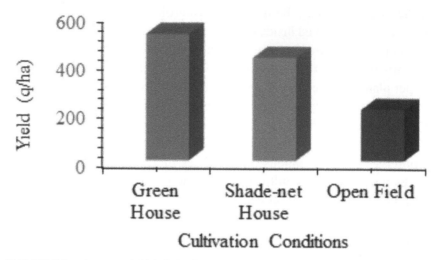

FIGURE 2.6   Average yield (q/ha) of capsicum in three conditions of cultivation, 1.0 q
= 100 kg.

house, respectively than the open field cultivation. Therefore, the study on
growth and yield of high value capsicum crop grown in three conditions
of cultivation gives an idea that high value crops are better to grow inside
protected structures to get maximum yield and quality produce, producing
approximately 2.5 and 2.0 times more yield in q/ha from greenhouse and
shade-net house cultivation, respectively than the open field cultivation.

2.3.2   COST AND ECONOMICS OF CAPSICUM UNDER DRIP IRRIGATION

The agricultural production economics of capsicum crop was estimated
considering the different components of cost of production and presented
in Table 2.9. The unit cost for each component were collected and used for
analysis of production cost. The cost of production does not include the
cost for erection of protected cultivation structures and the cost involved
in setting up of drip irrigation system. The experiment was conducted considering the existing systems of protected cultivation structures and drip
irrigation system only.

TABLE 2.9 Cost of Production of Capsicum in Three Conditions of Cultivation

| | Cost of production of capsicum in three conditions | | |
| | Greenhouse | Shade-net house | Open field |
Components	Indian Rupees, Rs.		
Land Preparation	35.00	40.00	40.00
Nursery/Seedlings	115.00	120.00	125.00
Manures and Fertilizers	250.00	250.00	250.00
Plant Protection	40.00	50.00	60.00
Hired Human Labor	550.00	550.00	560.00
Land Revenue	25.00	25.00	10.00
Total Cost for 20 m² Area	1,015.00	1,035.00	1,045.00
Total Cost (in Rs./ha)	**456,750.00**	**465,750.00**	**470,250.00**

Note: 1.00 US$ = 60.00 Rs.

Crop production data was also collected for the crop grown from October 2014 to June 2015. Among all the components that are responsible for the total cost of production of capsicum in three conditions of cultivation, the hired labor cost is maximum, followed by the cost and fertilizers of manures, plant protection measures and seedling nursery, etc., The total production cost analysis of capsicum crop for three different conditions of cultivations revealed that lowest production cost was required for growing crops inside the greenhouse than the shade-net house and open field because of high cost in weed control and plant protection measures. The cost of production for capsicum grown were 3% and 1% higher in open field (Rs. 470,250.00) cultivation than the greenhouse (Rs. 456,750.00) and shade-net house (Rs. 465,750.00) cultivation, respectively.

Economic analysis was carried out considering the investment, operation and production costs and the results are presented in Table 2.10. As the production cost values differed in different conditions of cultivation, the gross return (Rs./ha), net return per unit area (Rs./ha), net profit per unit production (Rs./100 kg) and B:C ratio values were found significantly higher in greenhouse followed by shade-net house and open field cultivation for capsicum crop. According to economical evaluation, considering the selling price (Rs./100 kg) same for three different conditions of cultivation for the capsicum crop, the highest return of Rs. 1,630,624.68 per hectare was found in greenhouse followed by shade-net house (Rs. 1,235,035.84) and open field (Rs. 377,249.33) cultivation. As compared to cost of cultivation of capsicum in three conditions of cultivation, net return was found highest from the greenhouse than shade-net house and open field. Therefore, farmers can get 4.32 times more net return for growing of capsicum in the existing greenhouse than the open field. Similarly, farmer can also get 3.27 times more net return for growing of capsicum in shade-net house than the open field. As compared to the net return for the production of capsicum in the protected cultivation structure and open field, it is suggested to grow capsicum inside the protected cultivation structures (greenhouse/shade net house) instead of growing in open field.

However, it was also noted that protected cultivation structures resulted significantly more income of capsicum crop because of high selling price and more yield of capsicum. Therefore, the benefit cost analysis may be the alternative to decide the feasibility of cropping pattern in different cultivation structures.

The benefit cost ratio (B: C Ratio) for the production of capsicum crop was 3.57, 2.65, and 0.80 for the greenhouse, shade-net house and open

TABLE 2.10 Economics of Production for Drip Irrigated Capsicum in Three Conditions of Cultivation

Particulars	Conditions of cultivations		
	Greenhouse	Shade-net House	Open Field
Cost of Cultivation (Rs./ha)	456,750.00	465,750.00	470,250.00
Average yield (q/ha)	521.84	425.20	211.87
Average cost of production (Rs./100 kg) — A	875.26	1095.38	2219.47
Average price received (Rs./100 kg)	4,000.00	4,000.00	4,000.00
Gross returns (Rs./ha)	2,087,374.68	1,700,785.84	847,499.33
Net returns per unit area (Rs./ha)	163,062,4.68	1,235,035.84	377,249.33
Net profit per unit production (Rs./100 kg) — B	3,124.74	2,904.62	1,780.53
B:C Ratio, C = B ÷ A	3.57	2.65	0.80

Note: 1.00 US$ = 60.00 Rs.

field, respectively. Thus, the protected structures resulted significantly higher benefit-cost ratio for growing of capsicum crop. Therefore, it is suggested to grow capsicum crop only inside the existing protected cultivation structures (greenhouse/shade net house) to increase the net annual income against the investment on existing infrastructure. The present study was conducted only by considering the existing set-up of greenhouse and shade-net house.

Authors also considered the existing drip irrigation system in all three conditions of cultivations (greenhouse, shade-net house and open field). However, in absence of existing protected structures and drip irrigation system, the payback period for set up of the similar kind of protected cultivation structures along with drip irrigation system needs to be estimated for the feasibility of the system in hilly terrain.

Payback period for capsicum crop was evaluated considering two components of the system such as protected cultivation structures and drip irrigation system and presented in Table 2.11. The life span of 12 years for

TABLE 2.11 Payback Period for Drip Irrigated Capsicum in Three Conditions of Cultivation

	Conditions of cultivation		
Components	Greenhouse	Shade-net House	Open Field
Protected Structure (Rs./m²)	1,970.00	840.00	0.00
Govt. Subsidy 60% for NEH Region (Rs./m²)	788.00	336.00	0.00
Drip Irrigation System (Rs./m²)	310.00	95.00	95.00
Govt. Subsidy 40% for NEH Region (Rs./m²)	186.00	57.00	57.00
Total Cost per unit Area (m²)	974.00	393.00	57.00
Total Cost in Rs./ha	9,740,000.00	3,930,000.00	105,000.00
Net returns per unit area (Rs./ha)	1,630,624.68	1,235,035.84	377,249.33
Net returns per Year (Rs./ha)	3,261,249.36	2,470,071.68	377,249.33
Life of Protected Structures (Years)	12	3	0
Life of Drip Irrigation System (Years)	15	12	10
Payback Period in years	2.99	1.59	0.28

Note: 1.00 US$ = 60.00 Rs.

greenhouse and 3 years shade-net house was considered for estimation of payback period of the system. And the life span of drip irrigation system for greenhouse, shade-net house and open field were considered 15, 12 and 10 years, respectively. In addition, there is provision of Govt. subsidy of 60% for protected structure and 40% for drip irrigation system on the total set up cost for North Eastern Hilly Region. Therefore, the subsidy amount was considered for the said analysis. It was also found that the capsicum could be grown two times in greenhouse and two times in shade net and only once in open field.

The results for payback period of capsicum cultivation in three conditions of cultivations revealed that the initial cost of erection of greenhouse with drip irrigation system is higher as compared to shade-net house and open field cultivation, as open field cultivation contains only the cost of drip irrigation system. The payback period for the greenhouse, shade-net house and open field condition for the production of capsicum was 2.99, 1.59 and 0.28 years, respectively. It revealed that the investment incurred for the setting up of protected cultivation structures (greenhouse and shade-net house) as well as drip irrigation system in case of open field only could be achieved within the life which shows the feasibility and more significant of the system. Also it was observed that the payback period among the three conditions of cultivation is not much for cultivation capsicum crop, because of high value cash crop, though the payback period is more in case of greenhouse followed by shade-net house and open field.

2.4   CONCLUSIONS

The present study on effect of three conditions such as greenhouse, shade net and open field cultivation of capsicum crop in hilly region conditions revealed that protected cultivation through low cost shade net and high cost greenhouse cultivation are the only alternative to increase the vegetable production round the year in the hilly terrain.

The results from field experiments on growth and yield of Capsicum under different conditions of cultivation with drip irrigation system revealed that protected cultivation structures are the only solution to high value crop with higher yield and better income. The higher yield of fruits could be contributed by the high-tech greenhouse structure which has got sufficient irrigation water through drip irrigation system, optimum temperature, radiation and humidity for plant growth and also provide protection from insects and pests. And more yields with less area and duration is also possible from small or marginal land of undulating north eastern hilly terrain with protected cultivation and drip irrigation system.

The agricultural production economic study revealed that though the initial cost involvement for cultivation in greenhouse is quite higher as compared to low cost shade-net house cultivation. As compared to cost of cultivation of capsicum in three conditions of cultivation, net return was found highest from the greenhouse than shade-net house and open field. And as compared to the net return for the production of capsicum in the protected cultivation structure and open field, it is suggested to grow capsicum inside the protected cultivation structures (greenhouse/shade net house) instead of growing in open field. However, it was also noted that protected cultivation structures resulted significantly more income of capsicum crop because of high selling price and more yield of capsicum. The protected cultivation structures resulted significantly good B:C ratio for growing of capsicum crop and thus it is suggested to grow capsicum crop only inside the existing protected cultivation structures (greenhouse/shade net house) to increase the net annual income against the investment on existing infrastructure.

The results for payback period of capsicum cultivation in three conditions of cultivations revealed that the initial cost of erection of greenhouse with drip irrigation system is higher as compared to shade-net house and open field cultivation, as open field cultivation contains only the cost of drip irrigation system. It revealed that the investment incurred for the setting up of protected cultivation structures (greenhouse and shade-net house) as well as drip irrigation system in case of open field only could be achieved within the life span which shows the feasibility and more significant of the system. Thus, protected cultivation with greenhouse could be used as the only one alternative to control the environment for maximizing crop productivity percent area and increasing the quality of vegetables produce year around in the hilly terrain of North East India. These studies can be extended to other parts of the world.

2.5   SUMMARY

The present studies are focused on the protected cultivation of capsicum crop in hilly region conditions. Field experiment was conducted to investigate the growth and yield of Capsicum under different conditions of cultivation with drip irrigation system. The results obtained from the aforementioned experimental study are summarized as under:

- Plant growth parameters such as the average height of plant, number of branches, leaves, fruits and yield of capsicum in three conditions of cultivation in hilly terrain were found as 192 cm, 11, 45, 14.03 and 1.30 kg, respectively for greenhouse; 87 cm, 10, 42, 12.04 and 1.06 kg, respectively for shade net house; and 45 cm, 8, 34, 40, 7.83 and 0.53 kg, respectively for the open field, respectively. Based on the trends of growth and yield parameters the summary revealed is that greenhouse gives comparatively better vegetative growth and quality of produce for capsicum than the shade-net house and open field cultivation.
- The average plant height inside greenhouse was recorded maximum followed by shade-net house and open field cultivation for the crop. The crop duration was also recorded maximum in case of Greenhouse cultivation for capsicum. Fruit nos. per plant, individual fruit weight and fruit sizes are also maximum for greenhouse cultivation, followed by shade-net house and open field cultivation.
- Considering the cultivation in greenhouse, shade-net house and open field, the yield of Capsicum in Greenhouse, Shade-net House and Open field cultivation could be of 521.84, 425.20 and 211.87 q/ha, respectively.
- The production cost of capsicum per hectare was found maximum for open field cultivation (Rs. 470,250.00) followed by shade-net house (Rs. 465,750.00) and greenhouse (Rs. 456,750.00) cultivation. Lowest production cost required for growing crops inside greenhouse than shade-net house and open field because of high cost involved in weed control and plant protection measures.
- The maximum net return per unit area (Rs./ha) for capsicum was Rs. 1,630,624.68 cultivated in greenhouse cultivation, compared to Rs. 1,235,035.84 in shade-net house and Rs. 377,249.33 in open

field, which gave the lowest net return among the three conditions of cultivation. Therefore, farmers can get 4.32 times more net return for growing of capsicum in the existing greenhouse than the open field, respectively. Similarly, 3.27 times more net return for growing of capsicum in shade-net house than the open field, respectively. The benefit cost ratio found for the production of capsicum crop was 3.57, 2.65 and 0.80, respectively for the greenhouse, shade-net house and open field.

- The payback period for the greenhouse, shade-net house and open field condition for the production of capsicum was found 2.99, 1.59 and 0.28 years, respectively.

Therefore, hi-technology of protected cultivation with greenhouse and shade net house, etc., are the only alternative for cultivation of vegetables year round in North-Eastern hilly terrain to alleviate extremes variation of rainfall, temperature and humidity, also the biotic stresses like photo stress, moisture stress, deficiencies in soil nutrients, excessive wind velocities and atmospheric carbon-dioxide and weeds growth in open field condition mainly during rainy and post rainy season.

KEYWORDS

- benefit cost ratio
- capsicum
- cost of cultivation
- drip irrigation
- economic analysis
- fruit size
- greenhouse
- gross return
- growth stage
- hi-tech poly house
- hillock
- initial cost

- **net return**
- **open field**
- **payback period**
- **plant height**
- **protected cultivation**
- **shade-net house**
- **yield**

REFERENCES

1. Anonymous (2005). A Comparative statement on cost of cultivation of annual horticultural crops under conventional and precision farming system in Tamil Nadu. www.tnau.ac.in/horcbe/tnpfp/economics.
2. Antle, J. M. (1999). The new economics of agriculture. *American Journal of Agricultural Economics, 81*, 993–1010.
3. Ayars, J. E., Phene, C. J., Hutmacher, R. B., Davis, K. R., Schoneman, R. A., Vail, S. S., & Mead, R. M. (1999). Subsurface drip irrigation of row crops: a review of 15 years of research at the Water Management Research Laboratory. *Agric. Water Manage., 42*, 1–27.
4. Barlow, C., & Jayasuriya, S. K. (1984). Problems of investment for technological advance: the case of Indonesian rubber smallholders. *Journal of Agricultural Economics, 35*, 85–95.
5. Batchelor, C. H., Lovell, C. J., & Murata, M., (1996). Simple micro irrigation techniques for improving irrigation efficiency on vegetable gardens. *Agric. Water Manage., 32*, 37–48.
6. Chakraborty, H., & Sethi, L. N. (2015). Prospects of Protected Cultivation of Vegetable Crops in North Eastern Hilly Region. *International Journal of Basic and Applied Biology (IJBAB), 2*(5), 284–289.
7. Chavas, J. P. (2001). Structural change in agricultural production: economics, technology and policy. Pages 263–285, In: Gardner, B., & Rausser, G. (eds.). *Handbook of Agricultural Economics, Volume 1.* Elsevier Science, Amsterdam, The Netherlands.
8. Chavas, J. P., Chambers, R. G., & Pope, R. D. (2010). Production economics and farm management: a century of contributions. *American Journal of Agricultural Economics, 92*, 356–375.
9. Dimitri, C., Effl, A., & Conklin, N. (2005). *The 20th Century Transformation of U.S. Agriculture and Farm Policy.* Economic Information Bulletin 3, Economic Research Service of the USDA, Washington D.C., USA.

10. Griliches, Z. (1963). The sources of measured productivity growth: United States agriculture, 1940–1960. *Journal of Political Economy, 71,* 331–346.
11. Hoppe, R. A., Korb, P., O'Donoghue, E. J., & Banker, D. E. (2007). *Structure and finances of U. S. farms: Family farm report.* Economic Information Bulletin 24, Economic Research Service of the USDA, Washington D.C., USA.
12. Karlberg, L., & Frits, W. T. P. V. (2004). Exploring potentials and constraints of low-cost drip irrigation with saline water in sub-Saharan Africa. *Phys. Chem. Earth, 29,* 1035–1042.
13. Kislev, Y., & Peterson, W. (1982). Prices, technology, and farm size. *Journal of Political Economy, 90,* 578–595.
14. Manchanda, A. K., Bhopal, S., & Singh, B. (1988). Effect of plant density on growth and fruit yield of bell pepper (Capsicum annuum L.). *Indian J. Agron., 33,* 445–447.
15. Maya, P., Natarajan, S., & Thamburaj, S. (1997). Effect of spacing, N and P on growth and yield of sweet pepper cv. California Wonder. *South Indian Hort., 45,* 16–18.
16. Nagarajan, M., Senthilvel, S., & Planysamy, D. (2002). Material substitution in Greenhouse construction. *Kisan World, 11,* 57–58.
17. Nair, R., & Barche, S. (2014). Protected cultivation of vegetables – present status and future prospects in India. *Indian Journal of Applied Research, 4(6),* 245–247.
18. Paul, C. M., Nehring, R., Banker, D., & Somwaru, A. (2004). Scale economies and efficiency in U.S. Agriculture: Are traditional farms history? *Journal of Productivity Analysis, 22,* 185–205.
19. Ruttan, V. W., & Binswanger, H. P. (1978). Induced innovation and the green revolution. Pages 358–408, In: Binswanger, H., & Ruttan, V. (eds.). *Induced Innovation: Technology, Institutions, and Development.* Johns Hopkins, Baltimore, USA.
20. Van Zanden, J. L. (1991). The first green revolution: the growth of production and productivity in European agriculture, 1870–1914. *Economic History Review, 44,* 215–239.

CHAPTER 3

WATER USE EFFICIENCY FOR DRIP IRRIGATED CHILI UNDER POLYETHYLENE MULCHING

M. M. DESHMUKH, M. U. KALE, S. B. WADATKAR, V. A. BHADANE, and S. M. GHAWADE

CONTENTS

3.1   INTRODUCTION

Chili (*Capsicum annum* L.) belongs to *Solanaceae* family and is classified as a vegetable crop. It is grown throughout the world under wide range of climatic conditions. It is grown almost throughout the country in India. The most important chili growing states in India are Andhra Pradesh, Maharashtra, Karnataka and Tamil Nadu, which together constitute nearly 75% of the total area [2].

It is necessary to make efficient use of water and bring more area under irrigation through available water resources, as the world becomes increasingly dependent on vegetable production. Drip irrigation under mulching provides potential for achieving moderate crop yields through improved water use efficiency and control of the soil environment, including water conservation [1]. Mulches are used for water conservation and erosion control in dry regions. Other reasons for high mulching use include: soil temperature modification, soil conservation, nutrient addition, improvement in soil structure, weed control and crop quality control [5]. Enhancing the population of natural enemies to manage pests of chili can be easily and effectively supplemented with cultural methods such as mulching but the suitable material like refractive silver/black plastic mulches helps to repel aphids and other insects, vector of viral diseases which damage plants [3, 6].

This chapter focuses on water use efficiency of drip irrigated chili crop.

3.2 MATERIAL AND METHODS

The field experiment on the effects of different irrigation methods on the crop growth and yield of chili was conducted during kharif season of 2014 at Chili and Vegetable Research Unit, Dr. Panjabrao Deshmukh Krishi Vidyapeeth, Akola, India. The field experiment was laid out in randomized block design, with four replications and five treatments of drip irrigation at different levels of evapotranspiration (ET) with and without plastic mulching.

T_1 – 40% ET with silver polyethylene mulch with drip fertigation
T_2 – 60% ET with silver polyethylene mulch with drip fertigation
T_3 – 80% ET with silver polyethylene mulch with drip fertigation
T_4 – 100% ET with silver polyethylene mulch with drip fertigation
T_5 – 100% ET without mulch with drip fertigation (control)

Before transplanting, common irrigation was applied on 20th July 2014 to bring the soil at the field capacity in each plot. Healthy seedlings of chili were transplanted on 21st July 2014 at a spacing of 60 cm (plant to plant) and 45 cm (row to row) in paired rows with distance of 75 cm between adjoining paired rows on broad bed furrow. The depth of water to be applied per plant was calculated by using Dick Krupp's formula given in following equation:

$$Q = A \times B \times C \times D \qquad (1)$$

where, Q = water requirement per plant (lit/plant); $A = ET_o = E_{pan} \times K_p$; B = crop coefficient ($K_c$); C = canopy factor; D = area allotted per plant (m²); E_{pan} = cumulative evaporation for two days; and K_p = pan coefficient (0.8).

Irrigation to chili crop was scheduled on every alternate day considering the cumulative pan evaporation of previous two days. In case of precipitation, it was cumulated for the same previous two days and cumulative rainfall subtracted from cumulative evaporation. If cumulative evaporation was more than cumulative rainfall, then remaining evaporation was taken for calculating the water requirement. If cumulative rainfall was more than cumulative evaporation, irrigation was not applied on that scheduled day. Moreover, irrigation was not applied for next two days due to excess rainfall than evaporation and considering the two days (48 hrs.) period for getting soil reached to its field capacity.

The water requirement, water saving, water use efficiency and yield of green chili were studied. Cost analysis was also worked out for each treatment.

3.3   RESULTS AND DISCUSSION

The field study was conducted to evaluate the effects of drip irrigation and plastic mulch on chili in terms of water saving, water use efficiency and yield of chili. The results are presented in Table 3.1.

Table 3.1 indicates that the water saving under drip irrigation system at 40, 60 and 80% ET levels with polyethylene mulching over 100% ET with and without polyethylene mulching was found to be 47.84, 31.89 and 15.95%, respectively. It is also indicated that by utilizing water equivalent to 100% ET with and without polyethylene mulching by adopting drip irrigation at 40, 60 and 80% ET with polyethylene mulching, the percent increase in irrigated area over 100% would be 92%, 47% and 19%, respectively. Considering the area of plantation and its plant population if one cannot irrigate his field with the requirement of 100 percent crop ET, then

TABLE 3.1 Comparative Statement of Water Use

Treatments	Water applied (ha-cm)	Water saving (percent)	Area would be irrigated by applying water equivalent to 100 % ET	Per cent increase in area over 100 % ET
T$_1$ (40% ET with PM)	22.50	47.84	1.92	92
T$_2$ (60% ET with PM)	29.38	31.89	1.47	47
T$_3$ (80% ET with PM)	36.26	15.95	1.19	19
T$_4$ (100% ET with PM)	43.14	–	–	–
T$_5$ (100% ET without PM)	43.14	–	–	–

irrigator can take the privilege to other treatments by changing the water requirement of suitable ET level.

3.3.1 YIELD OF GREEN CHILI AND WATER USE EFFICIENCY (WUE)

As the chili crop is vegetable crop, its harvesting was done from time to time by picking of fruits. The harvesting was completed in nine pickings. The data pertaining to average yield of green chili as influenced by poly-ethylene mulch and different irrigation levels is presented in Table 3.2.

Table 3.2 shows that treatment T$_3$ (80% ET + PM) recorded significantly highest yield of green chili (198.27 q/ha) and found at par with the treatment T$_4$ (100% ET + PM) and followed by treatment T$_2$ (60% ET + PM), T$_1$ (40% ET + PM). The lowest yield of chili was recorded in treatment T$_5$ (100% ET without PM) and was 103.66 q/ha at par with treatment T$_1$. It is also observed from the results that treatments T$_3$ and T$_4$ were at par and highest, but additional benefit in the treatment T$_3$ was less application of water, i.e., 15.95% saving of water. It seems that with less amount of water the higher and desirable yield can be obtained in the treatment T$_3$. These results are in conformity with those obtained by other investigators [4, 7, 8].

TABLE 3.2   Yield of Green Chili as Influenced by Polyethylene Mulch and Different Irrigation Levels

Treatments	Yield of green chili (100 kg/ha)	Consumptive use (ha-cm)	Water use efficiency (100 kg/ha-cm)	B:C ratio
T_1 (40% ET with PM)	117.81	22.50	5.24	1.02
T_2 (60% ET with PM)	147.09	29.38	5.01	1.27
T_3 (80% ET with PM)	198.27	36.26	5.47	1.71
T_4 (100% ET with PM)	187.39	43.14	4.34	1.61
T_5 (100% ET without PM)	103.66	43.14	2.40	1.08
F – test	Significant	–	–	–
SE (m) ±	8.899	–	–	–
CD at 5%	27.417	–	–	–
CV %	11.798	–	–	–

Note: One quintal, q = 100 k.g.

The higher growth and yields in the treatments of polyethylene mulch may be due to favorable moisture maintained in the root zone, its availability to plants, avoiding leaching of soluble fertilizer applied, no weed growth under polyethylene mulch (which avoids the competition in uptake of nutrients applied) and favorable environment maintained in rhizosphere.

Highest WUE was recorded in treatment T_3, which may be due to lowest water use, followed by treatments T_2, T_3, T_4. However, lowest WUE was recorded in treatment T_5. This may be due to the consumptive use in case of treatment of T_1 was lowest, whereas it was highest in case of treatment T_4 and T_5.

The maximum B:C ratio was obtained in treatment T_3 (1.71) followed by 1.61, 1.27, 1.08 and 1.02 for the treatments T_4, T_2, T_5 and T_1, respectively.

From the above results, it can be concluded that treatment T_3 (80% ET + PM) was found best among all the treatments. These results are in conformity with those obtained by [4] and [7]. The B:C ratio obtained in treatment T_5 (100% ET without PM) was found higher than that of treatment T_1 (40% ET + PM) due to lower cost of cultivation. Hence, the yields in treatment T_1 were higher than treatment T_5; and B:C ratio was found lower.

3.4   CONCLUSIONS

Drip fertigation with 80% ET replenishment under silver polyethylene mulch was found better in terms of growth, yield and B:C ratio for green chili production.

3.5   SUMMARY

An experiment was conducted to study the effect of mulching on water use efficiency of chili under drip fertigation at Dr. PDKV, Akola during July 2014 to February 2015. The experiment was laid out in randomized block design with five treatments which included four irrigation level (100%, 80%, 60% and 40% ET) with polyethylene mulch and (100% ET) without mulch replicating four times.

Higher plant growth, more number of fruits per plant and enhancement in the yield was observed in all treatments of drip irrigation with mulch. Yield of green chili was maximum in the treatment of drip irrigation at 80% ET with mulching (198.27 q/ha) and found to be at par with the treatment of drip irrigation at 100% ET with mulching (187.39 q/ha), It directly reflected 16% water saving with comparable yield. Minimum yield of chili was found in the treatment of drip irrigation at 100% ET without mulching (103.66 q/ha). Highest irrigation water use efficiency of 5.47 q/ha-cm was found in the treatment of 80% ET with mulching. On the basis of benefit cost ratio, it is economically viable for the farmers to adopt drip irrigation at 80% ET with mulching for green chili which shows BC ratio of 1.71.

KEYWORDS

- benefit cost ratio
- broad bed furrow
- canopy factor
- chili
- cost analysis
- cost of cultivation

- **cumulative evaporation**
- **drip irrigation**
- **evapotranspiration**
- **fertigation**
- **irrigation water use**
- **mulch**
- **pan coefficient**
- **polyethylene mulch**
- **randomized block design**
- **soluble fertilizer**
- **water requirement**
- **water use efficiency**
- **yield**

REFERENCES

1. Diaz, P. (2010). Bell pepper (*Capsicum annum* L.) grown on plastic film mulches: Effect on crop micro environment, physiological attributes, and fruit yield. *Hort. Science, 45*(8), 1196–1204.
2. Indian horticulture database, (2014). http://nhb.gov.in/area-pro/NHB_Database_2015.pdf, 6.
3. Ismail, S. M. (2012). Water use efficiency and bird pepper production as affected by deficit irrigation practice. *International J. Agril. Forestry, 2*(5), 262–267.
4. Pradhan, P. C., Mishra, J. N., Panda, S. C., & Behera, B. (2010). Technical feasibility and economic viability of drip irrigation and polyethylene mulch in mango (*Mangifera indica*). *Indian J. Dryland Agric. Res. Dev, 25*(2), 91–94.
5. Saroliya D. K., & Bhardwaj R. L. (2012). Effect of mulching on crop production under rainfed condition—A Review. *International Journal of Research in Chemistry and Environment, 2*(2), 8–20.
6. Sharma P. K., Sharma H. G., & Singh P. N. (2004). Drip irrigation and polymulches effect on growth and yield of sweet pepper – A Review. *Agric Rev., 25*(4), 304–308.
7. Singh S. R., Sharda P. P., Lubana S., & Singla C. (2011). Economic evaluation of drip irrigation system in bell pepper (*Capsicum annuum* L. var. Grossum). *Progressive Hort., 43*(2), 289–293.
8. Tiwari K. N., Mal P. K., Singh R. M., & Chattopadhyay A. (1998). Response of okra (*Abelmoschus esculentus* (L.) Moench.) to drip irrigation under mulch and non-mulch conditions. *Agril. Water Management*, 38, 91–102.

CHAPTER 4

PERFORMANCE OF WATERMELON UNDER MULCHING, SUBSURFACE AND SURFACE DRIP IRRIGATION SYSTEMS IN SEMI-ARID REGION

M. REDDY, M. S. AYYANAGOWDER, M. G. PATIL,
B. S. POLISGOWDAR, M. NEMICHANDRAPPA,
M. ANANTACHAR, and S. R. BALANAGOUDAR

CONTENTS

4.1   INTRODUCTION

Water plays an important role in crop production. Irrigation water is often limited and therefore the techniques which help to conserve water in the field are needed. Mulching is a recommended practice of moisture conservation in arid and semiarid regions.

Over the past decade, the use of plastic mulch in agriculture has emerged as a practice closely related to agricultural development in many developed countries. The agricultural and horticultural developments in USA, Western Europe, Israel and Japan have been made possible through extensive utilization of plastic mulching. The cultivation of high values crops using methods like drip irrigation, green house plastic much, etc., can give large income to small farmers.

Even with the rapid growth in production and use of plastics in India, the per capita consumption of plastics is only 2.2 kg which is very low as compared to consumption in developed countries like USA, Germany and Japan where per capita consumption is above 60 kg. World average of per capita consumption of plastic is 16.2 kg [5]. Sweet corn, tomatoes, cucumber, straw berry, lettuce, watermelon, okra, and grapes are the primary crops that are grown under plastic mulch.

The notable advantage of use of plastic mulch is its impermeability, which prevents direct evaporation of moisture from the soil and thus cuts down water losses [1]. Plastics like HDPE, LDPE, and LLDPE have been used as plastic mulch. Among these types of plastics, LDPE mulches are most commonly used. Recently LLDPE has been scoring over LDPE as a mulch material due to its two associated characteristics of better down gauging and puncture resistance, while checks weeds growth through it.

American Society of Agricultural Engineering (http://www.asabe.org) has defined subsurface drip irrigation as, "application of water below the soil surface through emitters, with discharge rates typically in the same range." At the beginning, "sub-irrigation" and "Subsurface irrigation" sometimes were referred for both SDI, and sub irrigation (water table management). "Drip/trickle irrigation" could include either surface or subsurface drip/trickle irrigation or both. SDI may also be defined as placement of drip pipe or hose along with drip lateral under specified depth so that normal mechanical operations could be carried out to ensure its use for several years [3, 4]. Subsurface drip irrigation has been successfully mostly used for the last 15–20 years efficiently. In this system mainline, sub-mainline, laterals and drip pipes are installed below the soil surface at specified depth (i.e., less than 12 cm deep).

This chapter discusses effects of mulching, surface drip irrigation and subsurface irrigation on performance of watermelon in the semi-arid region.

4.2   MATERIALS AND METHODS

During February 2014 to May 2014 and November 2014 to February 2015, the experiment was conducted at Main Agricultural Research Station, University of Agricultural Sciences (UAS), Raichur–India. The site was located at 16°15' N latitude, 77°20' E longitude and at an elevation of 389 m above mean sea level (MSL). The soil was clay loam in texture and had pH of 7.33.

There were three irrigation sub-treatments (80, 100 and 120% of ET in drip irrigation) and three main irrigation treatments (Surface drip irrigation with mulching, Surface drip irrigation without mulching, and subsurface drip irrigation), in a split plot design with four replications. Seedlings of watermelon (var. Sugar Queen) were transplanted at spacing of 2 m x 1 m The seedlings were transplanted in 36 beds of 10 m x 1 m (12 beds were drip with mulching, 12 beds were drip without mulching, and 12 beds were subsurface drip irrigation). One lateral of 16 mm diameter was used for each bed with an inline dripper at 90 cm distance and discharge of 4 lph. Irrigation was provided daily after calculating water requirement based on past 24 hours of pan evaporation.

4.3   RESULTS AND DISCUSSION

4.3.1   WATERMELON YIELD

Table 4.1 presents watermelon yield (tons per hectare) for mulch, without mulch and subsurface treatment of different irrigation levels during summer and winter seasons. During summer season (first season), the main plot with mulch gave maximum yield (65.75 tons) followed by subsurface (49.36 tons). The treatment without mulch recorded minimum yield (48.92 tons). Among the different irrigation levels, the plants receiving water at 80% ET gave maximum yield (57.50 tons) followed by 100% ET (55.38 tons). The lowest yield was noticed in 120% ET treatment (51.14 tons).

The interaction effects were significant. The treatment mulch with 80% ET recorded significantly maximum yield (71.18 tons) followed by 100% ET with mulch (65.28 tons). The significantly minimum yield was noticed in subsurface treatment of 120% ET (45.91 tons).

TABLE 4.1   Effects of Different Treatments on Yield (t ha^{-1}) of Watermelon

Treatment	During February 2014 to May 2014 (Summer)				During November 2014 to February 2015 (Winter)			
	I_1	I_2	I_3	Mean	I_1	I_2	I_3	Mean
	Yield (t ha^{-1}) of Watermelon							
T_1	71.18	65.28	60.78	65.75	70.72	64.50	59.71	64.97
T_2	48.28	51.73	46.74	48.92	47.76	50.64	45.70	48.03
T_3	53.03	49.13	45.91	49.36	52.03	48.76	45.02	48.60
Mean	57.50	55.38	51.14		56.83	54.63	50.14	
	SEM ±	CD at 5%			SEM ±		CD at 5%	
Main treatment	2.250	7.787			1.974		6.831	
Sub treatment	0.535	1.591			0.667		1.982	
I at same T	0.927	2.755			1.156		3.434	
T at the same or different I	2.492	7.404			2.383		7.080	

Main treatments: Sub treatments:

T_1: Mulch condition; I_1: Irrigation at 80% ET using drip irrigation.

T_2: Without Mulch condition; I_2: Irrigation at 100% ET using drip irrigation.

T_3: Subsurface drip irrigation; I_3: Irrigation at 120% ET using drip irrigation.

Similar trends were followed in winter season (second season) as shown in Table 4.1. The main plot with mulch recorded the maximum yield (64.97 t) followed by subsurface treatment (48.60 t). The treatment without mulch recorded the minimum yield (48.03 t). Among the different irrigation levels, the plants receiving water at 80% ET recorded maximum yield (56.83 t) followed by 100% ET (54.63 t). The lowest yield was noticed in 120% ET treatment (50.14 t).

The interaction effects were significant. The treatment mulch with 80% ET recorded the maximum yield (70.72 t) followed by 100% ET with mulch (64.50 t) which indicated significant differences with mulch and 120% ET (59.71 t). The minimum yield was noticed in subsurface treatment of 120% ET (45.02 t).

Combination of mulch with drip irrigation in different irrigation levels recorded the maximum yield than the subsurface and without mulch with drip irrigation plots. The Table 4.1 shows that plastic mulch with 80% of irrigation noticed the maximum yield (71.18 t ha^{-1} in summer season) and 70.72 t ha^{-1} in winter season). This was due to higher transpiration rate from the broader leaves even though plastic mulch reduces the evaporation from the soil. The present results obtained are in line with the findings of Tiwari et al. [7] and Vijay Kumar et al. [8].

4.3.2   AVERAGE FRUIT WEIGHT

Data pertaining to average fruit weight of both seasons is presented in Table 4.2. In first season it can be observed that the main plot treatment with mulch has recorded the highest average fruit weight (3.99 kg) followed by subsurface treatment (3.54 kg) and without mulch plot (3.54 kg). In the different levels of irrigation, the plant receiving water at 80% ET showed the highest average fruit weight (3.81 kg), which was on par with 100% ET (3.73 kg). The minimum average fruit weight was found in 120% ET (3.58 kg).

Among the interaction effected, the treatment with mulch and 80% ET has recorded the highest fruit weight (4.20 kg), which was on par with 100% ET with mulch treatment (3.95 kg). The lowest average fruit weight was recorded in 120% ET of subsurface treatment (3.45 kg).

In second season, Table 4.2 shows that the main plot treatment with mulch has recorded the highest average fruit weight (3.97 kg) followed by subsurface treatment (3.43 kg) and without mulch plot (3.39 kg). In the different levels of irrigation, the plant receiving water at 80% ET showed the highest average fruit weight (3.69 kg) which was on par with 100% ET (3.63 kg). The minimum average fruit weight was found in 120% ET (3.48 kg).

Among the interaction effects, the treatment mulch with 80% ET has recorded the highest fruit weight (4.15 kg), which was on par with 100% ET with mulch treatment (3.93 kg). The lowest average fruit weight was recorded in 120% ET of subsurface treatment (3.28 kg).

4.3.3   TOTAL SOLUBLE SOLIDS (TSS)

The effect of mulch, without mulch and subsurface drip irrigation with different irrigation levels on TSS of during seasons are presented Table 4.3.

TABLE 4.2   Effects of Different Treatments on Average Fruit Weight (kg)

	During February 2014 to May 2014 (Summer)				During November 2014 to February 2015 (Winter)			
	I_1	I_2	I_3	Mean	I_1	I_2	I_3	Mean
Treatment	Fruit weight (kg)							
T_1	4.20	3.95	3.83	3.99	4.15	3.93	3.83	3.97
T_2	3.53	3.63	3.48	3.54	3.35	3.53	3.30	3.39
T_3	3.70	3.60	3.45	3.58	3.58	3.45	3.28	3.43
Mean	3.81	3.73	3.58		3.69	3.63	3.47	
	SEM ±	CD at 5%			SEM ±		CD at 5%	
Main treatment	0.15	0.53			0.14		0.49	
Sub treatment	0.05	0.15			0.06		0.17	
I at same T	0.09	0.25			0.10		0.29	
T at the same or different I	0.18	0.54			0.18		0.54	

Main treatments: Sub treatments:

T_1: Mulch condition; I_1: Irrigation at 80% ET using drip irrigation.

T_2: Without Mulch condition; I_2: Irrigation at 100% ET using drip irrigation.

T_3: Subsurface drip irrigation; I_3: Irrigation at 120% ET using drip irrigation.

For first season, it can be seen that the treatment mulch showed the highest TSS value (14.40 brix) which was on par to subsurface (14.33 brix) and without mulch treatment (14.28 brix). In the sub plots, the irrigation at 120% ET recorded the maximum TSS (14.49 brix) and minimum TSS was found at 80% ET treatment (14.19 brix).

Among the interaction effects, the treatment mulch with 120% ET recorded the highest TSS (14.58 brix), which was on par with combination of mulch with 100% ET (14.38 brix) and mulch with 80% ET (14.25 brix). The minimum TSS was found in 80% ET without mulch (14.13 brix).

In second season, it can be seen that the treatment mulch showed the highest TSS value (14.35 brix), which was on par to subsurface (14.28 brix) and without mulch treatment (14.24 brix). In the sub plots, the irrigation at 120% ET recorded the maximum TSS (14.46 brix) and minimum TSS was found at 80% ET treatment (14.13 brix).

Among the interaction effects, the treatment mulch with 120% ET recorded the highest TSS (14.53 brix), which was on par with combina-

TABLE 4.3   Effects of Different Treatments on TSS (°Brix)

	During February 2014 to May 2014 (Summer)				During November 2014 to February 2015 (Winter)			
	I_1	I_2	I_3	Mean	I_1	I_2	I_3	Mean
Treatment	TSS, °Brix							
T_1	14.25	14.38	14.58	14.40	14.18	14.35	14.53	14.35
T_2	14.13	14.28	14.43	14.28	14.08	14.25	14.40	14.24
T_3	14.20	14.30	14.48	14.33	14.13	14.28	14.45	14.28
Mean	14.19	14.32	14.49		14.13	14.29	14.46	—
		SEM ±	CD at 5%		SEM ±		CD at 5%	
Main treatment		0.410	1.418		0.365		1.263	
Sub treatment		0.066	0.196		0.051		0.153	
T at same M		0.114	0.339		0.089		0.265	
M at the same or different T		0.430	1.279		0.379		1.127	

Main treatments: Sub treatments:

T_1: Mulch condition; I_1: Irrigation at 80% ET using drip irrigation.

T_2: Without Mulch condition; I_2: Irrigation at 100% ET using drip irrigation.

T_3: Subsurface drip irrigation; I_3: Irrigation at 120% ET using drip irrigation.

tion of mulch with 100% ET (14.35 brix) and mulch with 80% ET (14.18 brix). The minimum TSS was found in 80% ET without mulch (14.08 brix) condition.

4.3.4   CROP WATER PRODUCTION FUNCTIONS (CWPF)

The crop water production functions for watermelon were developed for different treatments such as mulch, without mulch and subsurface drip irrigation and different irrigation levels are 80, 100 and 120% ET. The maximum yield was recorded in mulch with all irrigation levels during both seasons. During first season, crop water use was 416.57 mm in 80%, 518.71 mm in 100% and 620.85 mm in 120% ET. In the second season, crop water use was 236.23 mm, 293.29 mm and 350.35 mm in 80%, 100% and 120% ET, respectively. The results are presented in Figures 4.1 and 4.2.

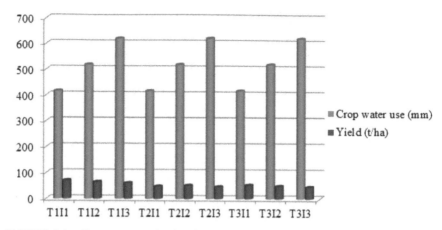

FIGURE 4.1   Crop water production functions during summer (First season).

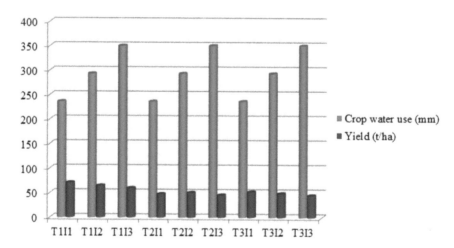

FIGURE 4.2   Crop water production functions during winter (Second season).

4.3.5 SOIL WATER DYNAMICS

The evolution of soil moisture storage during the growing season for selected soil depths of both seasons are presented in Figures 4.3–4.5. The results showed that there were greater variations in soil moisture storage in the surface soil layers compared with the deeper soil layers. The variation was more pronounced when depths (surface, 0.15 m and 0.30 m) were compared.

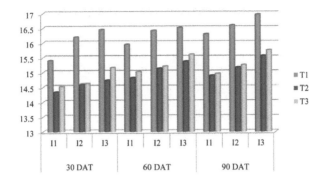

Soil moisture storage at surface level during summer: Y-axis, depth in mm

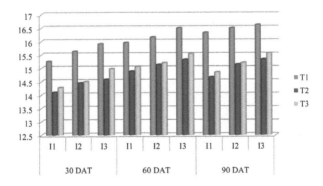

Soil moisture storage at surface level during winter, Y-axis, depth in mm

FIGURE 4.3 Soil moisture storage in selected surface level of soil layers for the 80, 100 and 120% ET of irrigation treatments (T1, T2, T3).

The pronounced variation in the surface layer of 0.15 m could be attributed to water uptake by plant roots, soil surface evaporation and drainage occurring in this zone. The intermittent wetting and drying of the soil profile caused high variation in the surface soil layers. Unlike in the surface soil layers, smaller variations were observed in the sub soil because the effective maximum rooting depth was 0.30 m.

This explains the smaller variations in the deeper soil layers because only fewer roots could reach this depth to extract soil water.

The interaction the combination of mulch, without mulch and subsurface with different levels of irrigation by 120% of ET shows the maximum soil moisture at all the irrigation methods. This was due to moisture distribution under drip irrigation is three dimensional function covering

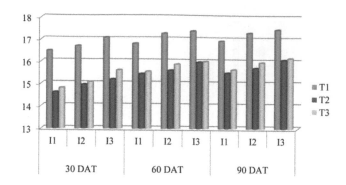

Soil moisture storage at 15 cm depth during summer: Y-axis, depth in mm

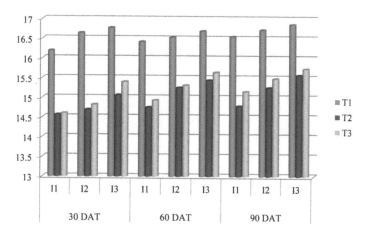

Soil moisture storage at 15 cm depth during winter: Y-axis, depth in mm

FIGURE 4.4   Soil moisture storage in selected 15 cm depth of soil layers for the 80, 100 and 120% of ET of irrigation treatments (T1, T2, T3).

vertical, lateral and diagonal movements, whereas it is a unidirectional movement under surface irrigation [2]. These results are in line with the findings of Raina et al. [6].

4.4   SUMMARY

The field experiment was conducted during 2014 and 2015 to assess the yield and yield traits of watermelon to evaluate effects of three main irrigation systems (like surface drip irrigation with mulching, surface drip

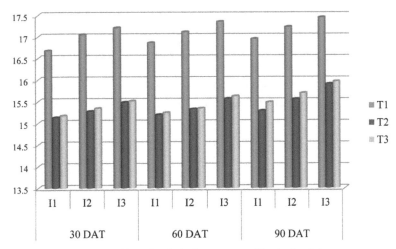

Soil moisture at 30 cm depth during summer: Y-axis, depth in mm

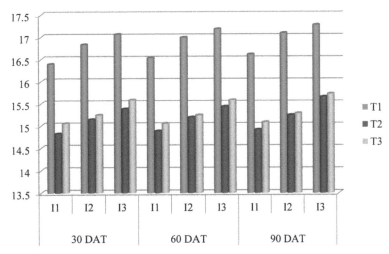

Soil moisture at 30 cm during winter: Y-axis, depth in mm

FIGURE 4.5 Soil moisture storage in selected 30 cm depth of soil layers for the 80, 100 and 120% ET of irrigation treatments (T1, T2, T3).

irrigation without mulching and subsurface drip irrigation)and three sub treatments (80, 100 and 120% ET) using drip irrigation

The yield varied from season to season. For summer season, water-melon yield varied from 71.18 t/ha (80% ET) of surface drip irrigation with mulching (T_1I_1) to 45.91 t/ha (120% ET) of subsurface drip irrigation (T_3I_3)

and same trend was followed in the winter season. The maximum average fruit weight was found in 80% ET (4.20 kg) of surface drip irrigation with mulching (T_1I_1) and the lowest average fruit weight was found in 120% ET (3.45 kg) with subsurface drip irrigation (T_3I_3) and same trend was followed in second season. In both seasons, highest yield was recorded in the 80% ET of surface drip irrigation with mulching than the other treatments.

KEYWORDS

- crop water production functions
- crop water use
- drip irrigation
- evapotranspiration
- high-density polyethylene
- linear low-density polyethylene
- low-density polyethylene
- moisture
- mulching
- subsurface
- subsurface drip irrigation
- surface drip irrigation
- total soluble salts
- watermelon

REFERENCES

1. Akbari, M., Dehghanisanij, H., & Mirlatifi, S. M. (2009). Impact of irrigation scheduling on agriculture water productivity. *Iranian J. Irrigation and Drainage, 1*, 69–79.
2. Badr, M. A. (2007). Spatial distribution of water and nutrients in root zone under surface and subsurface drip irrigation and cantaloupe yield. *World J. Agril. Sci., 3*(6), 747–756.
3. Megh R. Goyal (2015). *Research Advances in Sustainable Drip Irrigation*, volumes 1 to 10. Oakville, ON, Canada: Apple Academic Press, Inc. (http://appleacademicpress.com).

4. Megh R. Goyal (2017). *Innovations and Challenges in Sustainable Micro Irrigation,* volumes 1 to 8. Oakville, ON, Canada: Apple Academic Press, Inc. (http://appleacademicpress.com).
5. Iiyas, S. M. (2001). Present status of plastics in agriculture. Summer school on application of plastics in agriculture, 8–28 May, *CIPHET,* Ludhiana.
6. Raina, J. N., Thakur, B. C., & Bhandari, A. R. (1998). Effect of drip irrigation and plastic mulch on yield, quality, water use efficiency and benefit cost ratio of pea cultivation. *J. Indian Soc. Soil Sci., 46,* 562–567.
7. Tiwari, K. N., Ajai Singh, & Mal, P. K. (2003). Effect of drip on a yield of cabbage (*Brassica oleracea var. capitata* L.) crop under mulch and non-mulch condition. *Agril. Water Mgmt., 58,* 19–28.
8. Vijay Kumar, A., Chandra Mouli, G., Ramulu, V., & Avil Kumar, K. (2012). Effect of drip irrigation levels and mulches on growth, yield. *J. Res. ANGRAU., 40*(4), 73–74.

PART II

PERFORMANCE OF FRUIT
AND ROW CROPS

CHAPTER 5

DRIP IRRIGATION SCHEDULING OF *CITRUS RETICULATA* BLANCO (KINNOW): USING LOW COST PLANT LEAF TEMPERATURE SENSOR

M. DEBNATH, A. K. MISHRA, and N. PATEL

CONTENTS

5.1   INTRODUCTION

Agriculture is the major consumer of fresh water in the world. The widely practiced conventional irrigation techniques are less efficient where substantial amount of water is wasted. Water application efficiency can be increased significantly by selecting and designing the most appropriate method of irrigation suited for a specific location and crop. Irrigation by drip is very economical and efficient. In the conventional drip irrigation system, it is needed to keep a watch on irrigation timetable, which is different for different crops. Automated irrigation system can increase yields, save water usage, energy and labor costs as compared with manual system. Automatic irrigation systems presently available are more costly and are not adopted by most of the small and marginal farmers. Therefore, appropriate low cost technology has to be developed to facilitate high water use efficiency.

This chapter deals with a low cost automatic drip irrigation system for increasing water productivity and yield of fruit crops.

5.2   CHALLENGES WITH IRRIGATION SCHEDULING IN DEEP ROOTED CROPS

Irrigation scheduling in shallow rooted crops are easy as compared to deep-rooted crops. Shallow rooted crops are generally irrigated based upon soil moisture content measurement. Soil moisture sensors (gypsum blocks, nutron probe moisture meter, tensiometers etc.) are generally used for soil moisture content measurement also Time Domain Reflectrometry (TDR), and Frequency Domain Reflectrometry (FDR) are high cost soil moisture sensors for irrigation scheduling in shallow rooted crops. However in deep-rooted crops especially deep rooted horticultural crops, it is a difficult task to measure the soil moisture content in the effective root zone and effective irrigation scheduling is difficult to conduct in those crops.

5.3   PLANT SENSORS FOR DRIP IRRIGATION SCHEDULING

Irrigation scheduling based on plant water status rather than the soil moisture content is more accurate particularly in fruit crops having deeper root length making measurement of the soil moisture from deeper soil profile difficult. Studies on irrigation scheduling based on plant canopy temperature have been conducted by various researchers using sensors namely: infrared thermo-meter, infrared camera, thermocouple, and thermistor or by using multispectral remote sensing. These irrigation management techniques and instruments differ from each other with respect to their accuracy, labor intensity, cost and simplicity to use.

Beside soil moisture content, the plant water status also depends upon plant's surrounding atmospheric demand, plant rooting density and other plant characteristics [9]. Plant canopy temperature is an indicator of plant water stress and irrigation scheduling can be based on plant as an indicator of water stress [7, 8]. Plant canopy air temperature differential is a good indicator of plant water status [3]. It was found that the temperature of shaded leaf correlates better to plant water status [16]. For measurement of crop canopy temperature, contact type sensors such as thermocouple sensor or thermistor sensors are available [1, 2]. Non-contact type sensors such as infrared thermometer can also be used to measure the plant canopy temperature [11, 12]. However, the thermocouple sensor has disadvantage that its output signal is very small and is needed to be amplified to be used as a temperature sensor. The output of the thermistor sensor is nonlinear and also needed to be calibrated for getting the output in degree centigrade. The infrared sensors available for plant canopy temperature measurement are costlier and cannot be afforded by most of the farmers.

Irrigation can be controlled based on threshold canopy temperature [13]. Canopy temperature technique based on Temperature-Time Threshold (TTT) is also an option that can be used to schedule irrigation [10]. Microcontroller based system can efficiently be used to monitor crop temperature and water status [14]. Narrow (10°) field view wireless infrared sensor modules to measure plant canopy temperature have also been practiced in field [15]. LM-35-IC based microcontroller circuitry is a most recent, advanced leaf-air temperature differential measuring technology, which is best suited to automate a drip irrigation system in deep rooted fruit crops and thus saving irrigation water to increase crop yield remarkably [4].

5.4 MATERIALS AND METHODS

5.4.1 DRIP IRRIGATION AUTOMATION AND IRRIGATION SCHEDULING: USING LEAF-AIR TEMPERATURE DIFFERENTIAL SENSOR CIRCUITRY

The recently developed LM-35 IC based microcontroller circuitry is a low-cost, user friendly technology (Figures 5.1 and 5.2) to facilitate measurement of leaf-air temperature differential in deep rooted Kinnow crop

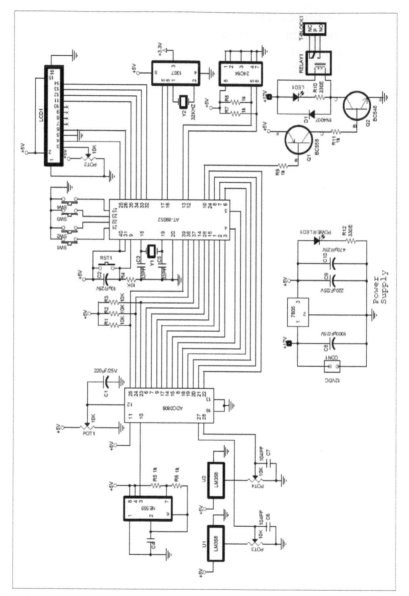

FIGURE 5.1   Designed microcontroller based leaf-air temperature differential circuitry.

[15]. The designed circuitry comprises, of: an IC 7805 Voltage Regulator, two LM 35 ICs, 555 integrated circuit, IC 24c64 memory chip, crystal oscillators, AT89S52 microcontroller, DS1307 RTC, an 16 x 2 LCD,

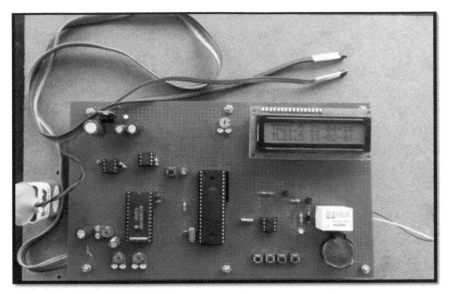

FIGURE 5.2   Developed leaf-air temperature differential sensor circuitry.

transistors BC558 and BC 548, micro switches, two LEDs, resistors including VR and capacitors.

The developed sensor circuit was installed in the field. One LM35IC was placed in contact at the lower side of the plant leaf and another nearer to the plant leaf to measure the ambient temperature. The sensors were placed in the shaded side of the plant during the noon time. The trends of variation in the leaf-air temperature differential in Kinnow crop was observed with the soil moisture content, air humidity, incident solar radiation and the leaf water potential. The leaf-air differential in the sunlit leaves was always higher than the shaded side leaves and it was always positive (i.e., in sunlit side, the leaf temperature was always higher than the surrounding air temperature). No trend was observed in the leaf temperature variation in the sunlit leaf with the soil moisture content and other climatic parameters. However, trend between leaf-air temperature differential of the shaded leaves with soil moisture content, humidity and leaf water potential was observed.

The temperature sensed by the two LM35 ICs used as the air-leaf temperature sensors converted into the digital values by the ADC converter enters to the micro controller unit. Time required for drip system operation and the irrigation interval was decided based on field observations and the conditions were set in the micro controller. The microcontroller unit is programmed so that the output pin of the microcontroller (pin no 10) remains in HIGH condition in normal situation, which indicates no output (0) from the output pin. When a desired leaf-air temperature differential condition reaches, the pin (pin no 10) becomes LOW and an output (1) is achieved which then activates the transistors assembly 558 and 548 and as a result the relay unit gets activated for the time period as decided by the micro controller unit. The output of the relay unit of microcontroller circuit passes through another two relay units one of which gives a signal to solenoid coil of the solenoid valve and another to pump starter unit through wire to run the pump automatically when the desired conditions fed into the microcontroller unit are reached thereby enabling automated drip irrigation.

Two step down transformers with rectifier circuits, one to power the sensor circuit with 12 V DC, 1 Amp current and another to power the solenoid valve with 24 V AC, 1 Amp current were used. The layout of the automated system is shown in Figure 5.3. An MCB DP as main power

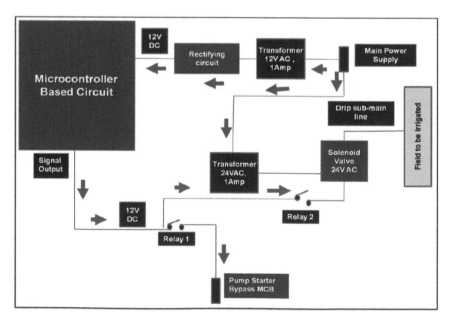

FIGURE 5.3   Layout of the leaf temperature differential based automated drip irrigation system.

supply switch was used to energize both of the transformers and a bypass MCB of pump starter was used to facilitate the manual operation of pump in unhealthy conditions.

5.5   RESULTS AND DISCUSSION

5.5.1   EFFECT OF PLANT SENSOR BASED DRIP IRRIGATION SCHEDULING ON WATER SAVING AND CROP YIELD

Plant canopy temperature based irrigation scheduling is most effective in case of deep rooted horticultural crops. Leaf-air temperature differential is an indirect indicator of plant water status, which is useful to save water in case of drip irrigation and also increase in crop productivity. The amount of water applied for Kinnow plant irrigated through conventional irrigation system (basin irrigation or manual drip irrigation) based on 100% of crop water requirement was found to be higher than that of the water applied by the developed system using microcontroller based leaf-air temperature differential sensor circuitry [6]. The water requirement of the Kinnow crop for manual drip irrigation and basin irrigation was calculated using following equations:

$$WR = (ET_c \times A_C \times K_C)/E \tag{1}$$

$$ET_c = ET_0 \times K_C \tag{2}$$

$$ET_0 = E_P \times K_P \tag{3}$$

where, WR is the water requirement of crop, m^3/day/plant; ET_c is the crop evapotranspiration; ET_0 is reference crop evapotranspiration; E_P is pan evaporation (USDA Class A Pan); K_P is the Pan coefficient in fraction (0.7); ET_c is the crop evapotranspiration, m/day; A_C is plant canopy area (m^2); and E is the efficiency of the irrigation system in fraction (0.90 for drip and 0.5 for basin irrigation).

Figure 5.4 shows water applied under manually operated drip irrigation system, basin irrigation system and the developed automated drip irrigation system in Kinnow crop. Table 5.1 shows the average crop canopy diameter, Leaf Area Index (LAI) and the amount of water applied under

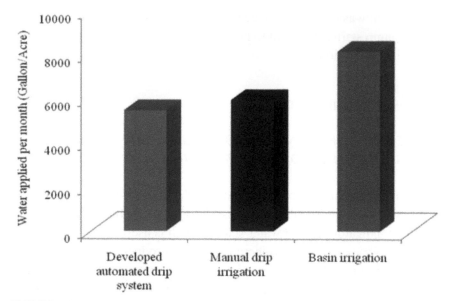

FIGURE 5.4 Water applied under different irrigation systems per month.

TABLE 5.1 Average Crop Canopy Diameter, Leaf Area Index (LAI) and Water Applied under the Conventional Drip System and the Developed Automated System

Parameters	Plants irrigated using developed automated system	Plants irrigated using manually operated drip system
LAI	1.55	1.57
Canopy diameter (ft)	5.30	5.31
Plant yield, lb/Acre	7,583.5	6,423.7

TABLE 5.2 Percentage Water Saving under Developed Automated System

| Parameter | Compared to | |
	Manually operated drip irrigation system	Basin irrigation system
Percentage water saving under developed auto-mated drip system	8.6	49.6

manually operated drip irrigation system and the developed automated drip irrigation system in Kinnow crop [16]. The average crop canopy diameter and leaf area index values for the plants irrigated under the developed automated system were found to be nearly equal to that of plants irrigated under conventional irrigation condition. It was observed that without affecting the plant growth, amount of irrigation water can be saved in sensor based automatic drip system. Table 5.2 shows the percentage water saving under the developed automated drip irrigation system as compared to manually operated drip irrigation and the basin irrigation system for Kinnow crop of 8 years old, spaced 16.4 ft × 16.4 ft [6].

The fully automated irrigation system includes all fundamental components and its price is about 180 US $.

5.6   CONCLUSIONS

Water is becoming a scarce commodity worldwide today. Low-cost automated drip irrigation system can be one of the best options to save irrigation water as well as increasing crop yield. Drip irrigation scheduling for shallow rooted crops can be conducted using soil moisture content measurement using available soil moisture sensors. However, irrigation scheduling for deep-rooted crops is a difficult task. Plant based measurements in case of deep rooted crops will be helpful for irrigation scheduling. Recently developed microcontroller based leaf-air temperature differential sensor circuitry for drip irrigation scheduling of Kinnow crop can also be used for various other deep rooted horticultural crops for irrigation scheduling thereby saving irrigation water and increasing crop yield. This study indicates that the developed low cost automated drip irrigation system based on plant leaf temperature sensor system can save 8.6% more water compared to manually operated drip system and 49.6% more water compared to basin irrigation system.

5.7   SUMMARY

This chapter deals with need of plant-based irrigation scheduling methods and their applicability in deep-rooted crops. This chapter explores potential of a recently developed low-cost user friendly plant sensor for drip irrigation scheduling in Kinnow crop and its applicability in other deep rooted fruit crops for saving of irrigation water and increasing yield of crop. The study reveals that considerable amount of irrigation water can be saved by using low cost automated drip irrigation system based on plant leaf temperature sensor system.

KEYWORDS

- automated drip irrigation
- basin irrigation
- canopy diameter
- crop coefficient
- crop evapotranspiration
- deep rooted crops
- deep rooted horticultural crops
- drip irrigation
- evapotranspiration
- irrigation scheduling
- leaf-air temperature differential sensor circuitry
- LM-35 IC
- microcontroller
- neutron probe
- pan coefficient
- plant yield
- reference crop evapotranspiration
- saving irrigation water
- shallow rooted crop
- soil moisture
- soil moisture content
- soil moisture sensor
- tensiometer
- water requirement

REFERENCES

1. Ahmed, M., & Misra, R. D. (1990). *Manual on Irrigation Agronomy*. Oxford and IBH Publishing Co. Pvt. Ltd., New Delhi, pp. 121–122, 272–282.
2. Abraham, N., Hema, P. S., Saritha, E. K., & Subramannian, S. (2010). Irrigation automation based on soil electrical conductivity and leaf temperature. *Agricultural Water Management, 45,* 145–157.
3. Bhosale, A. M., Jadhav, A. S., Bote, N. L., & Varsheya, M. C. (1996). Canopy temperature as an indicator for scheduling irrigation for wheat. *J. Maharashtra Agric. Univ., 21,* 106–109.
4. Debnath, M., & Patel, N. (2016). Performance of a developed low cost microcontroller based automated drip irrigation system in kinnow crop. In: *Innovation Energy Technology Systems and Environmental Concerns: A Sustainable Approach*, ISBN: 978-9384144-81-4.
5. Debnath, M., Patel, N., Mishra, A. K., & Varghese, C. (2016). Irrigation scheduling using low cost plant leaf temperature sensor based water application system for increasing water productivity of fruit crop. *International Journal of Electronics Communication and Computer Engineering, 7*(1), 2278–4209.
6. Debnath, M., Patel, N., Rajput, T. B. S., & Mondal, S. (2016). Water Saving under a developed microcontroller based automated drip irrigation system in Kinnow crop. *Journal of Soil & Water Conservation, 14*(4), 340–343.
7. Jackson, R. D. (1982). Canopy temperature and crop water stress. In: *Advances in Irrigation*, Vol. 1, Academic Press, New York, pp. 43–85.
8. Jacson, R. D., Idso, S. B., Reginato, R. J., & Pinter Jr. P. J. (1981). Canopy temperature as a crop water stress indicator. *Water Resour. Res., 17,* 1133–1138.
9. Kramer, P. J. (1969). *Plant and Soil Water Relationships: A Modern Synthesis*. McGraw Hill Book Co., New York.
10. Lamm, F. R., & Aiken, R. M. (2008). Comparison of temperature-time threshold- and et-based irrigation scheduling for corn production. *An ASABE Meeting Presentation* Paper Number 084202, Rhode Island Convention Center Providence, Rhode Island. St. Joseph, MI, pp. 207–213.
11. Mahana, J. R., & Yeater, K. M. (2008). Agricultural applications of a low-cost infrared thermomete. *Computers and Electronics in Agriculture, 64,* 262–267.
12. Saha, S. K. (1984). Remote sensing of crop evapotranspiration using plant canopy temperature. *Proceedings of the Seminar on Growth Condition and Remote Sensing*, IARI, New Delhi, India.
13. Steven R. E., Terry, A. H., Arland, D. S., Dan, R. U., & Don, F. W. (1996). Canopy Temperature Based Automatic Irrigation Control. *Proc. ASABE Annual International Conference*, San Antonio, TX.
14. Fisher, D. K., & Kebede, H. (2010). A low-cost microcontroller-based system to monitor crop temperature and water status. *Computers and Electronics in Agriculture, 74,* 168–173.
15. Susan, A. O., Martin, A. H., Steve, R. E., & Paul, D. C. (2011). Evaluation of a wireless infrared thermometer with a narrow field of view. *Computers and Electronics in Agriculture, 76,* 59–68.

16. Udompetaikul, V., Upadhyaya, S. K., Slaughter, D. C., & Lampinen, B. D. (2011). Plant water stress detection using leaf temperature and microclimatic information. *ASABE Annual International Meeting, Paper No. 1111555*, Louisville, Kentucky, St. Joseph, MI: ASABE.

CHAPTER 6

RESPONSE OF DRIP IRRIGATED BANANA TO DIFFERENT IRRIGATION REGIMES

S. K. PATTANAAIK

CONTENTS

6.1 INTRODUCTION

Water is life. It is the most important natural resource next to soil for crop production. The demand of water in agriculture is decreasing day-by-day. Hence, a sustainable management option and judicious use of water is present day challenge throughout the world. The switch over to horticultural crops and applications of micro-irrigation seems a promising proposition, when we look to the future with scarcity of water availability, nutritional security of nation and surplus food grain production [4, 9, 12]. Micro irrigation is the method of irrigation of application of small but frequent amount of water at the root zone. It eliminates the losses due to percolation and surface runoff. Thus, it gives very high application efficiency up to 90–95%.

India is the second largest producer of fruits after China. It produces about 44 million tons of fruits on a coverage area of 3.72 million hectares under fruits like: mango, guava, banana, lemon, citrus, pine apples, etc., Apart from these, other fruits like papaya, sapota, anola, jackfruits, berries, pomegranate, etc., that are also grown in tropical and sub-tropical climates. Banana is one of the most vital fruits grown in sizeable amount of area in almost all the states in India. It ranks second largest fruit crop in India occupying 13% of area and accounting for about 32% of total production of fruits. However, India ranks first position in the world in production of banana. The average production in the country is 32.5 ton/ha [10, 11].

Banana is one of the popular and remunerative fruit crop grown extensively in Odisha. In 1997–98, the banana production was 2,57,430 tons on an area of 23,896 ha with a productivity of 11.83 t/ha, which is still below the national average. The banana plant needs a good amount of water for high production and its yield is affected under deficit irrigation [1, 2]. In Odisha, different varieties of banana are grown under un-irrigated or marginally irrigated conditions while irrigation facility is essential for cultivating improved dwarf tissue culture variety such as Cv. *Dwarf Cavendish*. The prevailing irrigation practices for banana in Odisha are based mostly on experience of the farmers and not on experimental evidences. Hence, this recourse needs to be utilized in a judicious and scientific manner.

Drip irrigation, with its ability of small and frequent irrigation application, has created interest because of less water requirements, possible increased production and better quality of produce. Use of soil cover or mulching is also known be beneficial chiefly through their influence on soil moisture conservation, solarization and control of weeds. Information on the combine effects of drip with mulch on banana cultivation under Odisha condition has not been well documented.

The present experiment was planned to study the effects of different levels of irrigation on growth and yield of drip irrigated banana under mulch and non-mulch conditions.

6.2 MATERIALS AND METHODS

6.2.1 EXPERIMENTAL LAYOUT

A field experiment was laid out during 1998–1999 at Central Farm of Orissa University of Agriculture and Technology (OUAT), Bhubaneswar, located at 20°15'N latitude, 85°52'E longitude and at an altitude of 25.9 m above mean sea level (MSL). Soil in the study area is lateritic sandy loam having maximum water holding capacity of 31%, bulk density of 1.65 g/cm³ and infiltration rate of 12 mm/h. Tissue cultured plant-lets of cultivar Dwarf Cavendish were planted at a spacing of 1.5 m × 1.5 m. The recommended fertilizer doze and compost were applied per plant to meet the nutritional requirements of plant. Standard cultural practices were also followed as scheduled for banana cultivation. There were eight plants in each treatment and the experiment was laid out in a randomized block design in three replications with eight treatments as follows:

- T_1 = 100% irrigation requirement (V) through drip irrigation.
- T_2 = 80% irrigation requirement (0.8 V) through drip irrigation.
- T_3 = 60% irrigation requirement (0.6 V) through drip irrigation.
- T_4 = 100% irrigation requirement (V) by ring basin irrigation.
- T_5 = 100% irrigation requirement (V) through- drip irrigation with black LDPE (Low Density Poly Ethylene) mulch film.
- T_6 = 80% irrigation requirement (0.8 V) through drip irrigation with black LDPE mulch film.
- T_7 = 60% irrigation requirement (0.6 V) through drip irrigation with black LDPE mulch film.
- T_8 = 100% irrigation requirement (V) by ring basin irrigation with black LDPE mulch film.

Irrigation was initially given uniformly to all plants under all treatments up to 45 days after planting. Thereafter, black LDPE mulch film of 50 micron thickness was placed around the respective girth of each plant with 40% surface coverage [7, 8].

6.2.2 HYDRAULIC PERFORMANCE

In order to evaluate the performance of the drip system, the emitter discharge at different emitter locations in the laterals was recorded at operating pressures 0.4, 0.6, 0.8, and 1.0 kg/cm². Then mean emitter discharges, uniformity coefficient, and emitter flow variation were determined.

6.2.3 ESTIMATION OF IRRIGATION WATER REQUIREMENT (V)

Reference crop evapotranspiration (ET_0) was calculated using modified Penman method [6]. The crop coefficients (K_c) for different growth stages of banana were selected [5]. The actual crop evapotranspiration was estimated by multiplying reference crop evapotranspiration, crop coefficients, area under each plant and wetting fraction. The crop water requirement of banana crop was estimated by using the following equation [13].

$$V = (ET_0 \times K_c \times A_p) - (A_s \times R_e) \tag{1}$$

where, V = net volume of irrigation (liters/day/plant); ET_0 = reference crop evapotranspiration (mm/day); K_c = crop coefficient; A_p = A×W = effective area to be irrigated (m²); A = area allocated to each plant (m²); W = wetting fraction (0.3–0.5 for fruit crop); and R_e = effective rainfall (mm/day).

6.2.4 IRRIGATION SYSTEM

The water requirement was estimated for the growing season of banana during July to June. Daily time of operation of drip irrigation system was worked out. In drip, irrigation was scheduled on alternate days; hence total quantity of water delivered was cumulative water requirement of two days minus effective rainfall (if rain occurred). The lateral lines were laid along the crop rows and each lateral served each row of crop. The laterals were provided with 'on line' emitters of 4 lph discharge. Response of Banana to drip capacity was done in such a manner that water emitting out of emitter could wet the entire root zone of the plant. The duration of delivery of water to each treatment was controlled with the help of gate valve provided at the inlet end of each lateral. Figure 6.1 shows the arrangement of laterals and emitters in each treatment in field. Figure 6.2 shows the banana crop being irrigated by drip irrigation system.

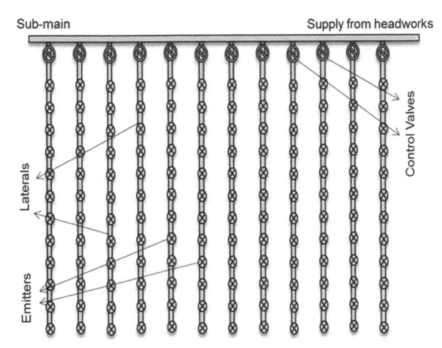

FIGURE 6.1   Layout of the experimental set up and banana plantation.

FIGURE 6.2   Drip irrigation system in banana plantation (Cv. *Dwarf Cavendish*).

In case of ring basin irrigation, irrigation was scheduled at weekly interval. The cumulative depth of water required for seven days was estimated and supplied to each plant.

6.3   RESULTS AND DISCUSSION

6.3.1   *EVALUATION OF HYDRAULICS PERFORMANCE OF DRIP IRRIGATION SYSTEM*

The observed data were analyzed following the procedure as outlined by Bralts et al. [3]. The optimum operating pressure was found to be 0.8 kg.cm/. At this pressure, the uniformity coefficient was maximum, i.e., 98.9% and the emitter flow variation was minimum, i.e., 10.75. The mean emitter flow was observed to be 4.23 lph, which is nearest to the designed emitter discharge of 4 lph and is within the permissible error of about 6%.

6.3.2   *EFFECTS OF DRIP IRRIGATION AND MULCHING ON YIELD*

There exists a significant influence of treatment of banana growth and yield (Table 6.1). Early emergence of flowers by fifty days was observed in case of plants treated with 0.8 V drip irrigation with black LDPE mulch (T_6) compared to 1.0 V ring basin without mulch film (T_4).

The drip irrigation in combination with mulch significantly increased the yield of banana compared to drip irrigation or ring basin irrigation without mulch. Among various treatments, the highest yield (525.33 qt/ha) was recorded under 0.8 V drip irrigation with mulch film (T_6). The yield was 52% higher than that in basin irrigation. Under drip irrigation with black LDPE mulch film for different levels of irrigation (T_5, T_6, T_7), percentage increase in yield were 14%, 30%, and 4%, respectively compared to drip irrigation without mulch (T_1, T_2, T_3).

The reasons of low yield in basin irrigated crops may be due to that the crop has to undergo water stress during last few days before next irrigation, especially at critical period, coupled with aeration problem during last few days immediately after irrigation. Moreover due to heavy appli-

TABLE 6.1 Effect of Different Treatments on Days to Flowering and Yield of Banana

Treatments	T_1 (V_D)	T_2 (0.8 V_D)	T_3 (0.6 V_D)	T_4 (V_B)	T_5 (M, V_D)	T_6 (M, 0.8 V_D)	T_7 (M, 0.6 V_D)	T_8 (M, V_B)
	Growth parameters							
1. No. of days to flowering	258.61	261.00	261.50	302.50	252.20	256.50	258.12	209.25
2. Yield (qt/ha)	408.00	404.44	374.23	345.34	464.45	525.33	387.78	368.44

Note: In this chapter, One US$ = 60.00 Rs. (Indian Rupees); 1.00 qt (quintal) = 100 kg.

cation of irrigation water, the nutrients may have got leached down the root zone of the crop. Also it may be attributed to high weed infestation between the crop rows.

In conventional basin irrigated method, although flowering and maturity started earlier (T_8) or (T_4) compared to the other treatments, it lagged behind in the maturity of the crop. The reason may be due to the moisture stress experienced between the irrigation intervals and competition of weeds for nutrient uptake by the crop.

6.3.3 COST-BENEFIT ANALYSIS

Table 6.2 presents the economic analysis of banana production under various treatments. The seasonal cost of cultivation included: expenses incurred on plowing, seedling, planting, inter-cultivation, de-suckering, application of fertilizer, planting, manuring, plant protection measures and laying of drip irrigation system.

The net seasonal income was found to be the highest for 0.8 V drip irrigation with black LDPE mulch film (T_6) compared to 1.0V basin irrigation without mulch (T_4). Gross benefit cost ratio was found to be the highest (2.8) for 0.8 V drip irrigation with mulch (T_6). Net profit in rupees per mm of water used, water use efficiency was found to be maximum for the 0.8 V drip irrigation with black LDPE mulch film (T_6).

6.4   CONCLUSIONS

The drip irrigation is economical and cost effective when compared with conventional basin irrigation. The use of drip either alone or with mulch can increase the yield of banana significantly over basin irrigation to the tune of 52%. To irrigate one hectare of banana crop with drip irrigation (1.0 V), 800 mm of water will be needed for sub-humid agro-climatic conditions of Bhubaneswar, Odisha, India. The duration of operation of drip irrigation is 18 minutes during initial growth stage which subsequently is increased to 66 minutes during peak demand of the crop with the emitter capacity of 4 lph discharge for each plant. The study also revealed that 52% increase in yield and about 54% higher net seasonal income could be

TABLE 6.2 Economic Analysis of Various Treatments

Cost economics	T_2 (0.8 V_D)	T_2 (0.8 V_D)	T_2 (0.6 V_D)	T_4 (V_B)	T_5 (M, V_D)	T_6 (M, 0.8 V_D)	T_7 (M, 0.6 V_D)	T_8 (M, V_B)
Note: In this chapter, One US\$ = 60.00 Rs. (Indian Rupees); 1.00 qt (quintal) = 100 kg								
1. Fixed cost (Rs)	70,000.00	70,000.00	70,000.00	-	70,000.00	70,000.00	70,000.00	-
(a) Life in years	7	7	7		7	7	7	
(b) Depreciation (Rs)	10,000.00	10,000.00	10,000.00	-	10,000.00	10,000.00	10,000.00	-
(c) Interest @ 9%	6,300.00	6,300.00	6,300.00	-	6,300.00	6,300.00	6,300.00	-
(d) Repair and maintenance @ 1%	700.00	700.00	700.00	-	700.00	700.00	700.00	-
(e) Total operational cost = [l(b)+ l(c)+ l(d)]	17,000.00	17,000.00	17,000.00	-	17,000.00	17,000.00	17,000.00	-
2. (a) Cost of Mulching	-	-	-	-	7,816.00	7,816.00	7,816.00	7,816.00
(b) Cost of cultivation	50,150.00	50,150.00	50,150.00	50,150.00	50,150.00	50,150.00	50,150.00	50,150.00
3. Total cost of cultivation per year = [l(e)+2(a)+2(b)]	67,150.00	67,150.00	67,150.00	50,150.00	74,966.00	74,966.00	74,966.00	57,966.00
4. Water used (mm)	800	640	480	800	800	640	480	800
5. Yield of produce (qt/ha)	408.00	404.00	374.22	345.33	464.44	525.33	387.78	368.44

TABLE 6.2 (Continued)

Cost economics	T_2 (0.8 V_D)	T_2 (0.8 V_D)	T_2 (0.6 V_D)	T_4 (V_B)	T_5 (M, V_D)	T_6 (M, 0.8 V_D)	T_7 (M, 0.6 V_D)	T_8 (M, V_B)
6. Selling price (Rs./ Kg)	4.00	4.00	4.00	4.00	4.00	4.00	4.00	4.00
7. Income from produce = [(5) × 100×(6)] in Rupees	1,63,200.00	1,61,776.00	1,49,688.00	1,38,132.00	1,85,776.00	2,10,132.00	1,55,112.00	1,47,376.00
8. Net seasonal income (Rs.) = [(7)–(3)]	96050.00	94626.00	82538.00	87982.00	110810.00	135166.00	80146.00	89410.00
9. Additional area cultivated due to water saved (ha)	–	0.25	0.665	–	–	0.25	0.665	–
10. Additional Expenditure due to additional area (Rs) = [(3)×(9)]	–	16787.5	44654.75	–	–	18741.5	49852.39	–
11. Additional income due to additional area (Rs) = [(7) × (9)]	–	40,444.00	99,542.52	–	–	52,533.00	1,03,149.48	–
12. Additional net income (Rs) = [(11)–(10)]	–	23,656.5	54,887.77	–	–	33,791.5	53,297.09	–

13. Gross cost of production (Rs) = [(3)+(10)]	67,150.00	83937.5	111804.75	50,150.00	74,966.00	93,707.50	1,24,818.39	57,966.00
14. Gross income (Rs) = [(7)+(11)]	1,63,200.00	2,02,220.00	2,49,230.52	1,38,132.00	1,85,776.00	2,62,665.00	2,58,261.48	1,47,376.00
15. Gross benefit cost, (B:C) ratio = [(14)/(13)]	2.43	2.41	2.23	2.75	2.47	2.80	2.06	2.54
16. Net profit per mm of water used (Rs./mm) = [(8)/(4)]	120.06	147.85	171.95	109.97	138.51	211.19	166.97	111.76
17. Water use efficiency in % (WUE)	51	63.19	77.96	43.16	58.05	82.08	80.78	46.05

obtained by using drip irrigation with black LDPE mulch film compared to conventional basin method without any mulch. Economic analysis also supports the findings that drip irrigation along with mulch is techno-economically feasible in the region to grow the crop.

6.5   SUMMARY

Banana is one of the popular fruit crops in Odisha. Farmers in the state generally irrigate the crop by ring and basin methods of irrigation that require large quantity of costly irrigation water. Water is becoming scarcer for irrigation in the state. Because of rapid industrialization and urbanization, water demand in non-agricultural sector is increasing day-by-day which compels share of water to agriculture sector to dwindle. Hence, it is imperative to adopt micro irrigation to save irrigation water and to enhance the water productivity. In the present study, eight different treatments with various irrigation levels in drip and basin irrigation and mulch and non-mulch conditions were tested in banana crop. The results indicate that the drip irrigation with black LDPE mulch significantly increases the yield of crop compared to non-mulch condition with ring basin method of irrigation. The highest yield of 525.33 (100 kg)/ha was recorded under 0.8 V drip irrigation with black LDPE mulch. This yield was 52% higher than the conventional method of ring basin irrigation without any mulch.

KEYWORDS

- **application efficiency**
- **banana**
- **basin irrigation**
- **benefit cost ratio**
- **cost-benefit analysis**
- **crop coefficient**
- **drip irrigation**

- effective rainfall
- emitter discharge
- fruit crop
- irrigation requirement
- laterals
- low density polyethylene
- micro irrigation
- moisture stress
- mulching
- net return
- operating pressure
- productivity
- randomized block design
- reference crop evapotranspiration
- replication
- root zone
- treatment
- wetting fraction
- yield

REFERENCES

1. Anonymous. (1990). *Training Program on Sprinkler and Drip Irrigation Systems.* Centre for Water Resources, College of Engineering. Anna University, Madras, July, 16–26.
2. Anonymous (2006). *Proceedings of the National Seminar on Micro-Irrigation Research in India: Status and Perspectives for 21st Century*, WTCER, Bhubaneswar, Orissa, July 27–28, pp. 17–29.
3. Bralts, V. F., Edwards, D. M., & Wu, I. P. (1987). Drip irrigation design and evaluation based on the statistical uniformity concept. *Advances in Irrigation, 4,* 67–117.
4. Chemmens, A. J. (1987). A statistical analysis of trickle irrigation uniformity. *Transactions of A.S.C.E., 30*(1), 169–173.
5. Doorenbos, J., & Kassam, K. (1983). Crop Water Stress on Yield. *FAO Paper No. 33,* Rome, Italy.

6. Doorenbos, J., & Pruitt, W. A. (1977). Guidelines for Predicting Crop Water Requirements. *FAO Irrigation and Drainage Paper No. 24*, Rome, Italy.

7. Panigrahi, B. (1992). Irrigation, mulches and irrigation scheduling in South Orissa. *Agricultural Engineering Today, 23*(3–4), 29–38.

8. Pattanaik, S. K. (2000). Response of Banana to Drip Irrigation with Mulching under Different Irrigation Regimes. *Unpublished MTech Thesis,* O.U.A.T., Bhubaneswar, Orissa.

9. Robinson, J. C., & Alberts, A. J. (1986). Growth and yield response of banana (cultivar 'Williams') to drip irrigation under drought and normal rainfall conditions in the sub-tropics. *Scientia Horticulture, 30,* 187–202.

10. Sivannappan, R. K. (1995). Increasing the production and profit through drip irrigation in tomato, a case study. *Proceedings of All India Seminar on Controlled Irrigation for Horticultural Orchard Crops of Arid Zones,* Institution of Engineers (India), Bhubaneswar.

11. Sivannappan, R. K. (2015). Micro irrigation potential in India. pp. 1, 87–120, In: *Research Advances in Sustainable Drip Irrigation.* Apple Academic Press, Inc.

12. Srivastava, P. K., Parikh, M. M., Saroni, N. G., & Raman, S. (1999). Response of banana to drip. *Proceedings of All India Seminar on Fruit Crops held at Institution of Engineers, India (Orissa State Centre, Bhubaneswar),* February.

13. Tiwari, K. N., Mal, P. K., Singh, R. M., & Chattopadhayay, A. (1998). Response of okra to drip. *Agril. Water Manag.,* 38, 91–102.

CHAPTER 7

HYDRAULIC PERFORMANCE OF LITCHI AND BANANA UNDER DRIP IRRIGATION

C. K. SAXENA, A. BAJPAI, A. K. NAYAK, S. K. PYASI, R. SINGH, and S. K. GUPTA

CONTENTS

7.1   INTRODUCTION

The future challenge to agriculture is to produce ever-increasing quantities of food and fiber not only with decreasing water availability for irrigation but also with availability of fresh irrigation water. Water is prime and the most precious natural resource as well as basic needs of life. Therefore, a sustainable management option and judicious use of water is present day challenge. The share of water for agriculture may reduce from present level of 84–69% by 2025 with increasing demand from the other sectors but on the other hand, demand of water for agricultural purposes would increase too. In the light of above challenges, switching over to horticultural crops coupled with micro irrigation could be a promising solution.

Micro irrigation works at a minimum or no losses in surface runoff and deep percolation at the same time it provides higher application efficiency generally around 80–90% or even higher. India is the second largest producer of fruits after China, with a production of 44.04 million tons of fruits from an area of 3.72 million hectares where mango, banana, citrus, guava, grape, pineapple and apple are the major fruit crops [1]. Apart from these, fruits like papaya, sapota, anola (Indian gooseberry), phalsa (*Grewia asiatica*), jackfruit, ber (*Ziziphus sp.*), pomegranate in tropical and sub-tropical group and peach, pear, almond, walnut, apricot and strawberry in the temperate group are also grown in a sizeable area. Most subtropical fruits are grown throughout India. The major fruit growing states are Maharashtra, Tamil Nadu, Karnataka, Andhra Pradesh, Bihar, Uttar Pradesh, and Gujarat.

Litchi (*Litchi chinensis* Sonn.) is one such crop, which is currently cultivated in an area of about 78,000 ha with a total production of around 497,000 tons [2]. The production of litchi is mainly concentrated in Bihar, West Bengal, Assam and Jharkhand and to a smaller extent in Tripura, Punjab, Uttarakhand, and Orissa.

India stands at the first position in the world in banana production. Banana (*Musa paradisiaca* L.) is grown in about 830.5 thousand ha with a total production of about 29.78 m tons. Banana comes next in rank occupying about 13% of the total area and accounting for about 32% of the total production of fruits. While Tamil Nadu leads other States with a share of 19%, Maharashtra has highest productivity of 58.60 tons/ha against India's average of 32.50 tons/ha [2]. The other major banana growing states are Karnataka, Gujarat, Andhra Pradesh and Assam. The main varieties of banana are Dwarf Cavendish, Bhusaval Keli, Basrai, Poovan, Harichhal, Nendran, Safedvelchi, etc. [1].

Drip irrigation is sometimes called trickle irrigation and involves dripping water onto the soil at very low rates (2–20 liters/hour) from a system of small diameter plastic pipes fitted with outlets called emitters or drippers. It is suitable for row crops (vegetables, soft fruits), tree and vine crops, etc., Knowledge of temporal hydraulic performance under a surface drip irrigation system is essential to analyze and evaluate the impact on crop yield over time, which call upon the service and maintenance of the system. The irrigation uniformity is an important indicator for such evaluation [8]. Besides the design of drip irrigation system, the knowledge of other drip hydraulic parameters such as the size of main line, sub main line and lateral and the capacity of used filtration and fertigation unit, the wetting front movement and the hydraulic performance depends upon the working condition of the system [11, 15].

In this context, a study was conducted to examine the utility of drip irrigation with cultivation of litchi and banana for evaluation of the temporal hydraulic variation in a drip irrigated fields.

7.2   MATERIALS AND METHODS

Two experiments were conducted in this study. The first one was conducted at the research farm of Central Soil Salinity Research Institute, Karnal located at 29°9'50"–29°50' N latitude and 76°31'15"–77° 12'45" E longitudes at an average altitude of 240 m MSL during 2002–05 in a field of 90 m x 48 m size. The soils of the experimental area belong to *Zarifa Viran* series of sub-order *Aquic Natrustalf*, is sandy loam to loam in texture at the surface (0–15 cm) and loam to clay loam in lower layers [12]. Soil characteristics consist of: average soil pH of the saturation paste 7.8, electrical conductivity of saturated paste extract, EC_e 0.7 dSm^{-1}, exchangeable sodium percentage, ESP 5.3, organic matter 2.9 g/kg soil, clay 15% [14]. A total of 130 litchi plants (*Litchi chinensis* Sonn. cultivar: Rose Scented) were planted in November 2000, the parameters in the study have been reported for 2004 and 2005. There were 13 laterals, each having 10 emitter locations for each plant, while each plant location had 4 emitter of 8 lph rating.

The second study was conducted at the Central Institute of Agricultural Engineering, Bhopal during 2013–2014 in a field of 30 m x 30 m size located at 77° 24' 10" E, 23° 18' 35" N at an average altitude of 495 m MSL. The soils of CIAE farm are vertisols having low infiltration rate of less than 10 mm/h. Soil characteristics consist of: average soil pH of the saturation paste 7.2, electrical conductivity of saturated paste extract, EC_e 0.3 dSm^{-1} and about 52% clay content [10]. A total of 144 banana plants (*Musa paradisiaca* L. cultivar: Dwarf Cavendish) were planted at a spacing of 2 m x 2 m distance on the field of about 900 m^2 in size during July 2012. The parameters in the study have been reported for the year 2013 and 2014. There were 12 laterals, each having 12 emitters of 8 lph rating for each plant. Figure 7.1 shows the general view of field experiments on litchi (Figure 7.1a) and banana (Figure 7.1b)

7.2.1 MEASUREMENT OF DISCHARGE

Discharge rate of individual emitters were computed by volumetric measurement over time using catch cans and stop watch. For the purpose of measurement of emitter discharge, the entire drip system was operated and allowed to run for about half an hour to stabilize and pressure fluctuations were not seen in the supply and lateral lines. The discharge rates for individual plant were obtained by summation of the discharges of all the emitters at a particular plant. The different rates of irrigation were obtained at various plant locations due to non-uniformity within the drip system. About 12% of emitters that were giving too high or too low discharge due to several reasons like clogging [9, 10] or damaged under normal wear and tear were replaced in the first year which improved the distribution of the discharge in particular case of litchi experiment, but the system remained intact in banana experiment.

| Fig. (a) | Fig. (b) |

FIGURE 7.1   General view of the experiments on litchi (a) and banana (b).

7.2.2 MEASUREMENT OF BIOMETRIC PARAMETERS

The plant height of the litchi and banana trees was measured from the soil surface to the highest point of the crown with the help of wooden scale and/or measuring tape. The girth at collar was converted from the mean diameter of the stem measured at about 5–10 cm above the soil surface with the help of Vernier caliper.

7.2.3 PERFORMANCE EVALUATION

The performance of the drip system for uniformity parameters was studied twice during the growth stages of both the litchi and banana field further it was correlated with the plant biometric parameters. The emitter discharge values were plotted using kriging technique through *Surfer* software to assess the variation in the irrigation within the experimental area and over two observations.

The arithmetic mean of plant height and girth at collar of the plants falling under similar discharge ranges were segregated and determined. For the litchi plants the emitter discharge range was selected from less than 10, 10–15, 15–20, 20–25, 25–30, and above 30 lph. While, for the banana plants emitter discharge ranged from less than 7, 7–7.5, 7.5–8, 8–8.5, and above 8.5 lph. Hydraulic performance of the drip system was studied by evaluating various uniformity parameters. Four uniformity parameters were computed using existing equations as described in the following subsections.

7.2.3.1 Christiansen's Coefficient of Uniformity

Christiansen [6] described the coefficient of uniformity as the ratio of absolute difference of each value from the mean and the mean. The Christiansen's Coefficient of Uniformity (CCU) can be expressed as:

$$CCU = \left[1 - \frac{\sum_{i=1}^{n}|D_i - \bar{D}|}{\sum_{i=1}^{n} D_i}\right]$$ (1)

where, D_i is the discharge or depth of irrigation of an emitter, D is the mean discharge of all emitters (or plants in case of plant wise determination of CCU), and n is total number of observations/emitters (total number of plants in case of plant wise determination of CCU).

This parameter considers deviations by magnitude alone without reflecting on excess or deficit. In practice, one of the two may be more critical.

7.2.3.2   Wilcox–Swailes Coefficient of Uniformity

Wilcox and Swailes [17] proposed a uniformity coefficient, Wilcox–Swailes Coefficient of uniformity (WSCU) based upon the coefficient of variation, which can be expressed as

$$WSCU = (1 - CV) \qquad (2)$$

where, CV is Coefficient of Variation expressed in fraction, as the standard deviation divided by mean value of emitter discharges (or the mean and the standard deviation of the sum of discharges of all emitters at each plant for plant wise WSCU).

This parameter has the same limitation as the CCU.

7.2.3.3   Statistical Uniformity

Hart [7] described the uniformity of irrigation using Statistical Coefficient of Uniformity (SCU) and Low Quarter Distribution Uniformity (SDU_{lq}), which are expressed as:

$$SCU = (1 - 2/\pi \; CV) \qquad (3)$$

$$SDU_{lq} = (1 - 1.27 \; CV) \qquad (4)$$

The reason for the use of term 1.27 in Eq. (4), as explained by Hart [7] is due to the fact that in a normal distribution, the mean of the low quarter of the values occurs approximately 1.27 times the standard deviation below the mean. These parameters have been used by many workers [3, 5, 13, 16]. While SCU has the same limitation as the CCU, SDU_{lq} reflects on the deficit of water in the lower quarter of the area if each dripper represents the same area.

7.2.3.4   Coefficient for Emitter Flow Variation

Bralts and Kesner [4] used Coefficient for Emitter Flow Variation (CEFV) that can be measured both plant and emitter wise from the field observations. It is expressed as:

$$CEFV = \frac{0.667 (\sum US - \sum LS)}{(\sum US + \sum LS)} \tag{5}$$

where, $\sum US$ is the sum of observations in upper 1/6th of distribution, and $\sum LS$ is the sum of observations in lower 1/6th of distribution.

Computed values of actual Coefficient of Uniformity (CU) were obtained from CEFV using the following equation described by Bralts et al. [4] as CU (CEFV).

$$CU(CEFV) = (1 - \sqrt{2}\ CEFV) \tag{6}$$

The values of the computed CUs are theoretically similar to that of CCU when the data follow a normal distribution. Therefore, a comparison between CCU and the computed values of CU (CEFV) were made to assess whether under our experimental set up CU could be calculated using CEFV at reduced cost on observations.

7.3   RESULTS AND DISCUSSION

The statistical parameters of the discharge at the plants are given in Table 7.1 over two years for both the litchi and banana plantation under micro irrigation. Since the values of the mean and the median are quite close while the values of standard deviation and coefficient of skewness are low, it indicates that the data follow a nearly normal distribution. It could be seen from Table 7.1 that the average plant height and girth at the collar were more at higher discharge rates and vice versa for banana experiment in the first year, whereas the same did not follow in the second year. This was mainly due to the fact that the most banana plants had attained their maturity by the age in the second year of observation and had attained similar biological biometric sizes, barring the range of discharge above 8.5 lph that had a relatively small representation out of total number of

TABLE 7.1 Statistical Parameters of Grouped Discharge (lph) in Litchi and Banana in Two Consecutive Years

Parameter	Litchi						Banana				
	Discharge at plant (lph)										
	<10	10–15	15–20	20–25	25–30	>30	<7	7–7.5	7.5–8.0	8–8.5	>8.5
First year											
Mean	8.25	12.73	17.57	22.31	26.86	34.66	6.41	7.36	7.82	8.25	10.92
Standard Error	0.91	0.33	0.23	0.24	0.29	1.01	0.32	0.03	0.02	0.03	2.15
Median	9.34	12.53	17.57	22.20	26.67	33.80	6.75	7.50	7.80	8.25	8.85
Standard Deviation	2.04	1.46	1.47	1.44	1.06	3.78	1.01	0.17	0.13	0.12	4.80
Sample Variance	4.15	2.12	2.15	2.08	1.13	14.31	1.01	0.03	0.02	0.01	23.01
Kurtosis	0.78	−0.84	−1.18	−1.00	0.44	−1.26	8.95	−0.68	−1.59	−1.26	5.00
Skewness	−1.30	−0.31	−0.01	0.17	0.54	0.39	−2.95	−0.88	−0.32	0.00	2.23
No of Plants	5	19	42	37	13	14	10	33	44	14	5
Maximum at 95 % Range	11.87	13.23	18.03	22.68	27.31	35.98	7.13	7.42	7.86	8.32	16.88
Minimum at 95 % Range	6.81	11.83	17.11	21.72	26.03	31.61	5.69	7.30	7.78	8.18	4.96
Mean Plant height (cm)	138.2	144.8	142.7	138.6	154.1	137.4	48.6	57.1	64.8	67.3	45.2
Mean Girth at Collar (cm)	9.9	11.2	11.1	10.9	10.8	10.0	30.7	33	35.9	38.4	27.2

Second year

Mean	7.74	12.98	17.71	22.72	27.37	34.35	4.67	7.23	7.73	8.17	10.23
Standard Error	0.00	0.41	0.24	0.29	0.26	1.08	0.40	0.06	0.03	0.03	0.46
Median	7.74	13.07	17.55	22.71	27.60	33.05	4.43	7.28	7.80	8.10	9.09
Standard Deviation	—	1.43	1.43	1.58	1.41	5.06	1.27	0.14	0.11	0.17	3.87
Sample Variance	—	2.04	2.05	2.49	1.98	25.61	1.61	0.02	0.01	0.03	14.99
Kurtosis	—	−1.40	−1.39	−1.23	−0.92	9.65	−0.01	−2.36	0.12	−1.36	13.90
Skewness	—	−0.14	0.14	−0.19	0.03	2.85	0.73	−0.44	−1.12	0.25	3.78
No of Plants	1	12	35	30	30	22	10	6	15	37	76
Maximum at 95 % Range	—	13.97	18.04	23.30	28.13	35.29	5.58	7.38	7.78	8.23	11.14
Minimum at 95 % Range	—	12.16	17.06	22.12	27.07	30.80	3.76	7.08	7.68	8.11	9.32
Mean Plant height (cm)	166.4	168.3	177.8	183.1	171.4	161.9	240.3	215.5	225.1	224.6	230.7
Mean Girth at Collar (cm)	13.8	15.1	14.6	15.8	14.3	13.6	46.1	36.1	44.4	44.8	44.5

plants. While in case of litchi plantation, it did not follow the similar trend for both the plant height as well as girth at collar. However, if we overlook the over irrigated range of above 30 lph, then the average values of the said parameters were higher in re-categorized range of above 20 lph than plant of lesser than 20 lph. Saxena and Gupta [13] had reported an increasing trend in some observations while no trend in some with emitter discharge, with segregation made over different pH levels from the same experiment.

The emitter discharge contours were drawn from 130 observed points in litchi and 144 observed points using the universal point kriging technique to assess the spatiotemporal distribution of plant wise rate of emitter discharge as observed during first and second years. The symbols in Figures 7.2(a,b) and 7.3(a,b) have been indicated as the plant locations where discharge were measured. Both of the contour maps displayed different patterns as the variation in discharge. The distribution of the discharge in litchi and banana field over two years could be seen and its variation could be assessed from the discharge contours in Figures 7.2 and 7.3, respectively. It can be seen from the both of the figures that the variability has gained in both of the experiment over time. It could be seen from the Figures 7.2(a) and 7.3(a) that the discharge was more uniformly distributed over the field as apparent from the less number of contours and there was lesser variability in discharge over both the fields. However, in Figures 7.2(b) and 7.3(b) the numbers of loops and contours were increased and it could be seen that the variability in emitter discharge rate has increased in the fields over the period of time.

The hydraulic performance of the system was assessed for the coefficient of uniformity computed using Equations 1 to 6 for two years and compared the temporal changes between two years. Table 7.2 enlists the overall scenario of emitter discharge and the uniformity parameters for both the experiments of litchi and banana for two years.

The values of mean, median and standard deviation were higher in second year than that of first year in both of the experiments. The major cause of a higher variation could mainly be due to the partial clogging in the systems. Although, the uniformity parameters for the litchi experiment are high in first year than second year mainly due to few changes of the emitters. It may be mentioned here that the system was amended as far as emitter discharge is concerned in the second year. However, in the banana

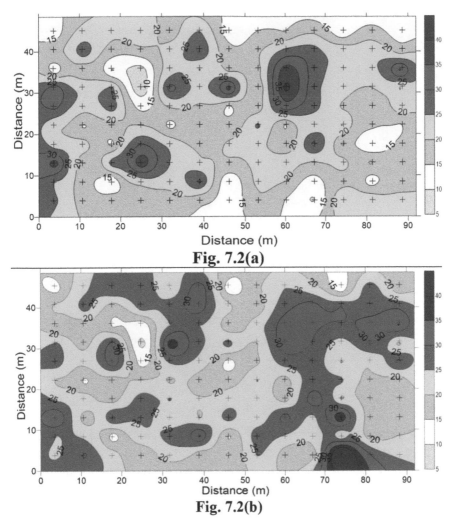

Fig. 7.2(a)

Fig. 7.2(b)

FIGURE 7.2 Emitter discharge (lph) contours for the year first year (a: top) and second year (b: bottom) in litchi field: Distances in the Y- and X- are in meters.

experiment, the uniformity reduced in second year. It could be observed from Table 7.2 that CUC, WSUC, SCU, SDUlq, CEFV and CU (CEFV) remained very good to excellent during the observational periods for the banana field.

Nearly close values of the CCU and SCU reveal that the statement made earlier that discharge data are normally distributed. As such, low

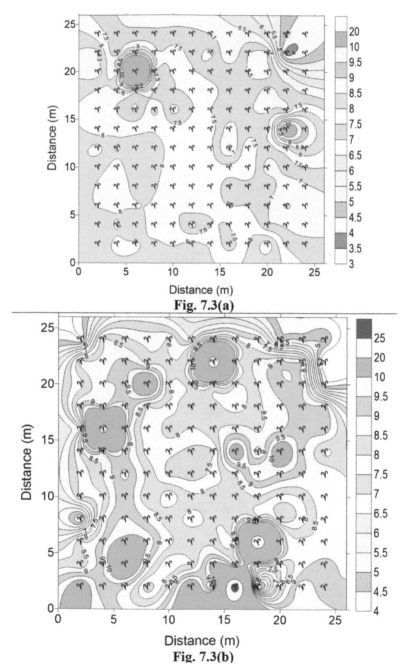

FIGURE 7.3   Emitter discharge (lph) contours for the first year (a: top) and second year (b: bottom) in banana field: Distances in the Y- and X- are in meters.

TABLE 7.2 Statistical and Uniformity Parameters of Discharge of Drip Irrigation Systems

Variable	Litchi		Banana	
	1st year	2nd year	1st year	2nd year
Statistical Parameters				
Mean	20.62	23.40	7.72	8.86
Standard Error	0.60	0.63	0.12	0.27
Median	19.78	23.18	7.65	8.48
Standard Deviation	6.89	7.13	1.30	3.18
Sample Variance	47.51	50.89	1.70	10.14
Kurtosis	0.66	1.41	56.97	21.76
Skewness	0.68	0.66	6.17	4.12
Number of observations	130	130	144	144
Maximum at 95 % Range	20.97	24.42	13.90	12.98
Minimum at 95 % Range	18.58	21.94	1.55	4.75
Uniformity Parameters				
Christiansen's Coefficient of uniformity (CCU)	0.74	0.76	0.93	0.84
Wilcox–Swailes Coefficient of Uniformity (WSCU)	0.67	0.70	0.83	0.64
Statistical Coefficient of Uniformity (SCU)	0.73	0.76	0.87	0.72
Low Quarter Distribution Uniformity (SDU_{lq})	0.58	0.61	0.79	0.55
Coefficient for Emitter Flow Variation (CEFV)	0.31	0.28	0.12	0.21
CU (CEVF)	0.75	0.78	0.91	0.83

WSCE values are expected and could be ignored for our study [13]. Ascough and Kiker [3] reported that the CU values (expressed in per cent) for various irrigation systems varied from 17.4 to 95.2% and it was close to 81.2% for drip and micro spray. The experiments under the study with a CU of 0.74 – 0.93 confirmed that both systems performed well. The values of SDU_{lq} were improved considerably after the change of emitters in litchi experiment which revealed that at least 25% of the area suffered due to low application, which improved after the changeover. This explains the reason why the decision to change the drippers was implemented. On the other hand it reduced in banana experiment over time with no change in emitter settings, the SDU_{lq} got down in the second year from 0.79 to 0.55. However, under the field conditions, a SDU_{lq} of around 0.70 is considered quite good [13].

Since the experimental data are normally distributed, a comparison of the CCU and the equivalent uniformity coefficient computed from the CEFV seems justified. The uniformity coefficient computed using CEFV i.e., UC(CEFV) remained closer to the CCU values. As such, for large systems the CCU could be computed through the evaluation of CEFV. This means, only a part of the system could be evaluated at low cost to compute CEFV and to arrive at the CCU of the system.

The CV (Coefficient of Variation) values computed from these data have remained above 33% during first year to 30% in second year in the litchi experiment. It could be seen the variability could be checked in over first year witch change of emitter points as stated in litchi experiment. The emitter discharge on two months of April 2013 and April 2014 has been significantly different (= 18.766).

In case of banana, the CV remained above 17% during first year to 36% in second year. It could be revealed that the value of CV for the first year reaming above 20 percent, it was indicated that considerable variability existed (Corwin et al., 2006). The emitter discharge for two years remained significantly different (= 13.582).

7.4   SUMMARY

The hydraulic performance of drip irrigation systems was evaluated for two years with litchi and banana crops, which were found to be good to excellent as revealed by various uniformity parameters such as Christiansen's Coefficient of Uniformity, Wilcox–Swailes Coefficient of Uniformity [17], Statistical Coefficient of Uniformity, Low Quarter Distribution Uniformity and Coefficient for Emitter Flow Variation. However, a careful watch is required during operation and maintenance so that performance remains high. The simplified technique of CEFV could be used to assess the coefficient of uniformity for the full system by evaluating only a part of the system.

- The Christiansen's Uniformity Coefficient (CUC) of the trickle irrigation system in litchi field was 0.74 and 0.76 whereas for banana field it was 0.93 and 0.84 for the first and second years, respectively. Similarly, the Wilcox–Swailes Uniformity Coefficient (WSUC), and Statistical Coefficient of Uniformity (SCU) for litchi and banana were found to be 0.67–0.70, 0.83–0.64; and 0.73–0.76, and 0.87–0.72 respectively for the first and second years. While the Low Quarter Distribution Uniformity (SDU_{lq}) improved from 0.58 to 0.61 in banana, while it reduced from 0.79 to 0.55 in a year's lapse, respectively.
- The emitter discharge parameters of both the experiment had shown temporal variability significant at 1 per cent level.
- The variability of emitter discharge was reduced over time was mainly due to the replacement emitters with extreme low or high discharge. The uniformity coefficient for emitter discharge computed using CEFV i.e., UC(CEFV) of 0.75 and 0.78 remained closer to the CCU values. While, the variability of emitter discharge among the banana experiment has increased. The uniformity coefficient for emitter discharge computed using CEFV i.e., UC(CEFV) of 0.91 and 0.83 against the CCU values of 0.93 and 0.84.

ACKNOWLEDGMENTS

Authors are thankful to the Director, CIAE, Bhopal, and the Director, CSSRI, Karnal for providing necessary funds, facilities and permissions and the Dean, CAE, JNKVV, Jabalpur for necessary permissions.

KEYWORDS

- *Aquic Natrustalf*
- banana
- Bhopal

REFERENCES

1. Agricoop (2014). *Horticulture (Heralding a Golden Revolution)*. Accessed: April 2014 http://agricoop.nic.in/hort/ hortrevo5.htm

2. APEDA 2016. *Litchi*. Agricultural & Processed Food Products Export Development Authority (APEDA), Ministry of Commerce & Industry, Govt. of India, NCUI Building 3, Siri Institutional Area, August Kranti Marg, New Delhi – 110016. accessed: Feb. (http://agriexchange.apeda.gov.in/Market%20Profile/one/LITCHI.aspx).

3. Ascough, G. W., & Kiker, G. A. (2002). The effect of irrigation uniformity on irrigation water requirements. *Water SA, 28*, 235–241.

4. Bralts, V. F., & Kesner, C. D. (1983). Drip irrigation field uniformity estimation. *Transactions of the American Society of Agricultural Engineers*, 26, 1369–1374.

5. Burt, C. M., Clemens, A. J., Sterlkoff, T. S., Solomon, K. H., Bliesner, R. D., Hardy, L. A., Howell, T. A., & Eisenhauer, D. E. (1997). Irrigation performance measures: Efficiency and uniformity. *Journal of Irrigation and Drainage Engineering (ASCE), 123*, 423–442.

6. Christiansen, J. E. (1942). Hydraulics of sprinkling systems for irrigation. *Trans. ASCE, 107*, 221–239.

7. Hart, W. E. (1961). Overhead irrigation pattern parameters. *Agricultural Engineering, 42*, 354–355.

8. Kishore, Ravi, Gahlot, V. K., & Saxena, C. K. (2016). Pressure Compensated Micro Sprinklers: A Review. *International Journal of Engineering Research and Technology, 5*(1), 237–242.

9. Pandey, R. S., Batra, L., Qadar, A., Saxena, C. K., Gupta, S. K., Joshi, P. K., & Singh G. B. (2010). Emitters and filters performance for sewage water reuse with drip irrigation. *Journal of Soil Salinity and Water Quality, 2*(2), 91–94.

10. Pandey, S. (2014). Dynamics of wetting front movement and its parametric relationship in subsurface point source of trickle in vertisols. Paper presented in the 48th ISAE Annual Convention & Symposium organized at College of Technology and Engineering (CTAE), Maharana Pratap University of Agriculture and Technology, (MPUAT), Udaipur during February 21–23.

11. Saxena, C. K., & Gupta, S. K. (2004). Drip Irrigation for Water Conservation and Saline/Sodic Environments in India: A Review. In: *Proceedings of International Conference on Emerging Technologies in Agricultural and Food Engineering*. IIT, Kharagpur. December 14–17. Natural Resources Engineering and Management and Agro-Environmental Engineering. Amaya Publishers. New Delhi, pp. 234–241.

12. Saxena, C. K., & Gupta, S. K. (2006a). Effect of soil pH on the establishment of litchi (*Litchi chinensis* Sonn.) plants in an alkali environment. *Indian Journal of Agricultural Sciences, 76*(9), 547–549.

13. Saxena, C. K., & Gupta, S. K. (2006b). Uniformity of water application under drip irrigation in litchi plantation and impact of pH on its growth in partially reclaimed alkali soil. *Journal of Agricultural Engineering, 43*(3), 1–9.

14. Saxena, C. K., Gupta, S. K., Purohit, R. C., Bhakar, S. R., & Upadhyay, B. (2013). Performance of okra under drip irrigation with saline water. *ISAE Journal of Agricultural Engineering, 50*(4), 72–75.

15. Saxena, C. K., Gupta, S. K., Purohit, R. C., & Bhakar, S. R. (2015). Salt water dynamics under point source of drip irrigation. *Indian Journal of Agricultural Research*, *19*(2), 101–113.

16. Solomon, K. H. (1984). Yield related Interpretations of irrigation uniformity and efficiency measures. *Irrigation Science*, *5*, 161–172.

17. Wilcox, J. C., & Swailes, G. E. (1947). Analysis of surface irrigation efficiency. *Sci Agriculture*, *27*, 565–583.

CHAPTER 8

MICRO IRRIGATION PRACTICES IN RICE GROWN UNDER THE SRI METHOD

D. MAHAPATRA, N. SAHOO, and B. PANIGRAHI

CONTENTS

8.1   INTRODUCTION

Food security is the state of having reliable access to a sufficient quantity of affordable food grains which depends on the ability of the farmer to increase production with reducing availability of water to grow crops.

This chapter is an edited version of Dinaranjan Mahapatra, "Effect of micro irrigation practices on yield and yield attributes of rice grown under SRI method." 2015. Unpublished thesis for Master of Technology, Department of Soil and Water Conservation Engineering, College of Agricultural Engineering and Technology, Orissa University of Agriculture and Technology (OUAT), Bhubaneswar, Odisha, India.

Rice as a staple crop, which was previously misjudged as a submerged crop, is a prime target for water conservation because it is the most widely grown of all crops under irrigation. In producing 1 kg of rice, farmers have to supply 2–3 times more water in the rice fields than other cereals. In Asia, there is a requirement of approximately 80% of the developed fresh-water resources for irrigation purposes, about half of which is used for rice production. Rapidly depleting water resources threaten the sustainability of the irrigated rice and hence the food security and livelihood of rice producers and consumers. In Asia, 17 million hectares (Mha) of irrigated rice areas may experience physical water scarcity and 22 Mha may have economic water scarcity by 2025. There is also much evidence that water scarcity already prevails in rice-growing areas, where rice farmers need technologies to cope with water shortage and ways and means must be sought out to grow rice with lesser amount of available water.

Rice is grown in three seasons in India: *kharif* (15th June to 15th October), *rabi* (25th October to 15th February) and summer (20th February to 31st May). The *kharif* season accounts for 88%, *rabi* and summer season accounts for 12% of total production of rice. In India, the rice crop is highly dependent on the South-West monsoon, which occurs over the sub-continent from 15th June to 15th October. Green revolution in India (1967–1978) brought substantial increase in production of cereals, particularly wheat and rice during the tenure of Late Prime Minister Indira Gandhi. Among the cereals, rice and wheat continue to dominate not only in India but also in Asia. These crops are grown in vast regions of the country due to its adaptability to a wider range of agro-climatic conditions. Thus, rice is the principal food grain for the present and also for the future> Hence management of rice crop production can emerge as the key area of management in the field of agriculture.

Rice cultivation in lowland ecosystem is very water demanding. Statistics indicate that the water consumption of rice accounts for approximately 54% of the total agricultural water consumption. It is estimated that 5000 liters of water is needed to produce 1 kg of rice. In addition to it, the rapid industrial and urban development demands for increase in fresh water, thereby leading to fresh water shortage. Thus, reducing agricultural water consumption is the only way left for rice crop to manage with the scarcity of water resources. The percentage of global paddy output for different countries is shown in Table 8.1 [1].

TABLE 8.1   Paddy Output for Some Leading Countries of the World

Country	Per cent share of global paddy output
India	21.5
Indonesia	8.6
Bangladesh	5.2
Vietnam	4.2
Myanmar	3.6
Thailand	3.5
Japan	2.7
Brazil	2.0
USA	1.7
Korea	1.3

Among the paddy growing countries, India has the largest area under cultivation, though in terms of volume of output, it is second to China (Table 8.1). Productivity in India is much lower than in Egypt, Japan, China, Vietnam, USA and Indonesia and even below the world's average. It makes up 42% of India's total food grain production and 45% of the total cereal produced in the country. In India, rice is the most preferred staple food as it is consumed by 65% of the population. It continues to play a vital role in the country's exports—constituting nearly 25% of the total agricultural exports from the country. One-third of the world's paddy cultivation area (83 million hectares) is in India. It is grown in almost all states of India but is mostly concentrated in the river valleys, deltas and low-lying coastal areas of north-eastern and southern India. The paddy producing states are: Assam, West Bengal, Bihar, Madhya Pradesh, Odisha, Andhra Pradesh, Tamil Nadu, Kerala, Karnataka, Maharashtra, Gujarat, Uttar Pradesh and Jammu and Kashmir, which together contribute over 95% of the country's crop. Of these, West Bengal, Odisha, Andhra Pradesh, Tamil Nadu and Bihar are the major contributors.

Indian share in global rice production has been hovering in the range of 19.50 to 24.52% as shown in the Table 8.1. Indian share dipped below 20% only in 2009–2010. Production of rice in India is expected to drop the year 2014 from 105.24 million tons (MT) to 103.13 MT due to low *kharif* output.

The artificial application of water to the soil is known as irrigation. It is used to assist in the growing of agricultural crops, protecting plants against frost, suppressing weed growth in grain fields and preventing soil consolidation. Various types of irrigation techniques differ in how the water obtained from the source is distributed within the field. In general, the goal is to supply water to the entire field uniformly, so that each plant gets due amount of water according to its need. In drip irrigation system, water is applied drop by drop just at the root zone of the plant. The drip irrigation method may be the most water-efficient method of irrigation, if managed properly. The field water use efficiency of drip irrigation is typically in the range of 90–95% when managed efficiently.

In modern agriculture, drip irrigation is often combined with plastic mulch to further reduce evaporation loss and is also the means of delivering soluble fertilizer effectively at the base of the plant. Drip irrigation methods range from very high-tech with computerized control to low-tech with manual control and is very labor-intensive. Low soil water tensions are usually provided to each plant in the field in comparison to other micro irrigation methods and the system may also be designed for achieving the highest water uniformity throughout the field in a landscape containing a mix of plant species instead of same plant species. Although it is difficult to regulate pressure on steep slopes in other micro irrigation systems, yet the pressure can be regulated more precisely with the help of pressure compensating emitters available in the market, so the undulating field never needs to be leveled. High-tech drip system is such a system which requires precisely calibrated emitters located along lines of lateral and these are controlled by computer operated solenoid valves.

Existing water-saving technologies for rice cultivation can be divided into three categories according to their water-saving capacity. The categories in sequence are:

- Continuously saturated soil cultivation system: It maintains high soil water content in all the growth stages, so water losses are high.
- The rice intensification system is known as "aerobic rice," in which, upland rice is grown under non-flooded condition with adequate inputs and supplementary irrigation by drip when rainfall is

insufficient to meet the consumptive demand of the crop. Because of a great reduction in seepage, percolation and evaporation, this technology allows for a greater Water Use Efficiency (WUE) and higher water saving compared to traditional flooded irrigation.

• The alternate wetting and drying system: It is ground cover rice production system (GCRPS) which gives higher water productivity and higher crop yield obtained under plastic mulching with drip irrigation than under furrow and sprinkler irrigation.

However, studies on water-saving technologies in rice production systems have mainly focused on the innovations in cultivation systems that incorporate furrow irrigation, sprinkler irrigation, drip irrigation and drip irrigation under plastic mulching which gave some inspiring records of water-saving for rice cultivation. In recent times, it is very important to increase the productivity, water use efficiency (WUE) and production of rice crop under the drip irrigation system under water scarce situation. Most of the farmers of the state and country cannot leave rice cultivation in upland ecosystem under water scarce situation resulting from scanty rainfall. Because cultivation of rice in *kharif* season is very much linked to their rich culture and tradition since time immemorial, therefore under such situation drip irrigation under plastic mulch is perhaps good option to grow rice in water scarce upland ecosystem of the state and the country.

The micro irrigation in general and drip irrigation in particular has received considerable attention from policy makers, researchers, economists, etc., for its perceived ability to contribute significantly to ground water resources development, agricultural productivity, economic growth and environmental sustainability. The drip method of irrigation has been found to have a significant impact on reducing the extensive use of fresh water, resources saving, cost of cultivation, yield of crops and farm profitability. Hence, the policy should be focused on promotion of drip irrigation in those regions where scarcity of water and labor are at alarming stage and where, shift towards wider-spaced crops is taking place.

The conventional rice cultivation method under irrigated conditions involves land leveling and construction of irrigation and drainage channels. During land preparation, rice seeds are soaked, ahead of planting. The sowing of soaking seeds is mechanized in some developed countries. In most developing countries, including India and China, the process is manual, with the seeds sown directly in the field along the line or on the nursery beds for transplanting. In the later case, traditionally seedlings are transplanted in bunches consisting of two to three seedlings of 21–30 days duration. Now there is a trend of direct seeding in the field surpassing transplantation, which has become a customary to reduce the duration of crop. The average plant population in that directly sown rice crop is roughly consisting of 200–300 seedlings per square meter. Traditionally, the rice fields are inundated with water for about three months or so, and are drained only before the harvesting. Currently, main stream technological options to improve rice production focus mainly on selection of improved varieties, proper crop nutrition and weed management, pest and diseases control and irrigation management. Interestingly, System of Rice Intensification (SRI) is an alternative technique that depends on such agronomic practices with less input requirement and maximizes yield.

The best part is that by adopting SRI on large scale, the government can save a considerable amount of money by avoiding expensive additional water supply structures. It also helps in reducing water conflicts and related costs to society and individuals, improving access to water for the poor and ensures sustainable ecosystems.

The research study in this chapter is focused to make an attempt to eradicate the issues regarding the production of paddy in water scarcity areas. Keeping the above facts in view, a field study was conducted with the following objectives:

- To estimate water requirement of rice crop under various irrigation practices.
- To compare yield and yield attributes of rice at different irrigation levels.
- To study the economics of different micro irrigation systems on rice crop.

8.2   REVIEW OF LITERATURE

8.2.1   DRIP IRRIGATION

Drip irrigation system delivers water to the crop root zone directly using a network of pipelines with emitters spaced along the length of the lateral based on crop geometry. Each emitter supplies a measured quantity of water, nutrients, hormones, pesticides and insecticides directly into the root zone of the crop at a precise and uniform rate for the required growth and protection of plants. Water and nutrients enter the soil from the tip of emitters, move into different layers of root zone of the plants through the combined forces of gravity and capillary. So the moisture and nutrient level in crop root zone is maintained at optimum level, ensuring that the plant never suffers from water stress and nutrient deficiency, thus enhancing quality and quantity. Drip Irrigation can be considered as a sustainable novel technology that allows rice production to maintain or increase in the face of declining water availability. Some of the relevant research papers on drip irrigation were reviewed and are described here in brief.

It has been concluded that drip irrigation systems allowed water to be applied uniformly and slowly at the base of the plant so that all the applied water is stored in the root zone [4]. Researchers reported that drip irrigation uses 30–50% less water than surface irrigation, reduces salinization and water logging and achieves up to 95% irrigation efficiency. Many researchers have observed that drip irrigation can be successfully used in commercial fields without increasing root-zone soil salinity, potentially eliminating the need for subsurface drainage-water disposal facilities [9, 13].

One of the demand management mechanisms is the adoption of micro irrigation such as drip and sprinkler methods of irrigation. Evidences showed that the water use efficiency (WUE) can increase up to 100% in a properly designed and managed drip irrigation system [13, 15].

Drip method of irrigation helps to reduce the over-exploitation of groundwater that partly occurs because of inefficient use of water under surface method of irrigation. Environmental problems associated with the surface method of irrigation like water logging and salinity are also completely absent under drip method of irrigation [16].

Developing infrastructure for the water resources and their management have been the common policy agenda in many developing countries, particularly in the arid and semi-arid tropical countries like India. A study by the International Water Management Institute (IWMI) revealed that around 50% of the increase in demand for water by the year 2025 can be met by increasing the effectiveness of irrigation. Drip method helps in achieving saving of irrigation water, increased water-use efficiency, decreased tillage requirement, higher quality products, increased crop yields and higher fertilizer-use efficiency [18].

Adoption of micro irrigation systems is likely to pick up fast in the arid and semi-arid areas, where farmers have independent irrigation sources, and groundwater is scarce. Further, large size farms and individual plots, and a cropping system dominated by widely-spaced row crops, which are also high-valued, would provide the ideal environment for the same [9]. A study at University of California concluded that a salt balance must be maintained in the root zone for productive cropping systems to continue, and irrigation without improved management practices cannot be sustained in the San Joaquin Valley. The options available to address salinity and drainage problems without retiring land are: (i) reducing drainage through better management of irrigation water; (ii) increasing the use of shallow groundwater for crop irrigation without any yield reductions; and (iii) reusing drainage water. All three methods require adequate salinity control in the root zone. As a result, drip irrigation is commonly used in salt-affected soils for producing vegetables [9].

Drip irrigation can apply water both precisely and uniformly at a high irrigation frequency compared with furrow and sprinkler irrigation, thus potentially increasing yield, reducing subsurface drainage, providing better salinity control and better disease management since only the soil is wetted whereas the leaf surface stays dry [17].

Drip irrigation at 150% of pan evaporation with drip fertigation of 100% recommended fertilizer dose, azophosmet and humic acid recorded 19% increased yield as compared to drip irrigation at 125% of pan evaporation with drip fertigation of 100% recommended fertilizer dose through drip. The increase in rice grain yield with drip irrigation at 150% of pan evaporation with drip fertigation, azophosmet and humic acid was mainly attributed by greater and consistent availability of soil moisture and nutrients which resulted in better crop growth, yield components and ultimately reflected on the grain yield [2, 14].

In a field study, it was found that highest grain yield and maximum water use efficiency [16] were also found in case of drip irrigated rice transplanted with 10 day old seedling spaced at 25 cm x 25 cm. over other treatments. Highest water productivity of 0.66 kg/m^3 was obtained under treatment four and lowest with conventional practice of rice cultivation with 0.37 kg/m^3. The total quantity of supplemental irrigation provided through drip irrigation was 291.42 mm whereas in conventional practice an amount of 553.3 mm was applied, which indicates not only saving in water but also the electricity consumption by about 58% [16].

8.2.2 SYSTEM OF RICE INTENSIFICATION (SRI) METHOD

System of Rice Intensification (SRI) is defined by as a technique of agronomic manipulation. The practices are based on a number of sound agronomic principles in order to achieve higher grain yield. The SRI is a method of rice cultivation, which is being practiced in more than 40 countries throughout the World. The SRI involves cultivating rice with as much organic manure as possible, starting with young seedlings planted singly at wider spacing in a square pattern; and with intermittent irrigation that keeps the soil moist but not inundated, and frequent inter cultivation with weeded that actively aerates the soil. SRI is not a standardized, fixed technological method. It is rather a set of ideas and methodology for comprehensively managing and conserving resources by changing the way that land, seeds, water, nutrients, and human labor are used to increase the productivity.

SRI is a productive set of practices, each proven, which can individually contribute to plant growth and development. SRI trials in Africa have shown that the yield potential of existing rice varieties can be doubled, without increasing agrochemical inputs and with saving of water. This is economically viable and environment friendly [19]. Researchers have found that one conventional rice plant produces 5 panicles whereas one SRI plant produces 8–10 panicles. Each conventional panicle contains 100–120 full grains while each SRI panicle has 180–200 full grains. It rightly responds to the pressures of high input costs and low margins in this tough business where many farmers have suffered. They heavily applied chemical fertilizers, thus, soil becomes unfertile. The overuse and abuse of herbicide spray makes the rice plants become unhealthy and more susceptible to diseases and less productive [3,6].

SRI represents an integrated and agro-ecologically sound approach to irrigated rice cultivation, which offers new opportunities for location specific production systems of small farmers. SRI is a designer innovation that efficiently uses scarce land, labor, capital and water resources, protects soil and groundwater from chemical pollution, and is more accessible to poor farmers than input dependent technologies that require capital and logistical support. SRI methods can lead to superior phenotypes and agronomic performance for a diverse range of rice genotypes [8, 10]. It improves physiological activities of the plant and provides better environmental condition. The key to success with SRI is the early transplanting of seedlings (8 to 12 days seedling), single planting with wider space 25–35 cm plant-to-plant and row-to-row.

In SRI method, extensive root systems and the improved structure and biological condition of soil were achieved by application of compost, which provided access to much larger pool of nutrients. The advantages from using compost have been seen from factorial trials, but if organic matter is not available, SRI practices can be also used successfully with chemical fertilizer. SRI is not a package of fixed technical specifications. It is rather a system of production formulated on certain core principles of soil chemistry and biology, rice physiology and genetics and the principles of sustainability with the possibility of adjusting the exact technical components based on the prevailing biophysical and socio economic realities of an area. This definition calls for research and adoption of the system to specific conditions of an area rather than trying to impose practices relevant to one location on the other injudiciously [4, 5].

A survey was conducted in Madagascar to investigate farmer implementation of AWD as part of SRI and showed that farmers have adapted AWD practices to fit the soil type, availability of water and labor. The primary drawbacks reported by farmers with implementing AWD were the lack of a reliable water source, little water control, and water-use conflicts. They suggested that by combining AWD with SRI, farmers can increase grain yields while reducing irrigation water demand. Methane released from agricultural activities largely comes from inundated rice fields and ruminant animals, which together produce almost half of human-induced methane. Methane is produced by anaerobic microbes in soils that are deprived of oxygen by continuous flooding. Making paddy soils intermittently and mostly aerobic substantially reduces methane emissions [12].

Average yield in case of SRI method increase was 78% (3.3 t/ha) with a 40% reduction in water use and 50% in fertilizer applications, with 20%

lower costs of production. SRI practices can achieve significantly higher output with a reduction in inputs, enhancing simultaneously the productivity of the land, labor, water and capital used in irrigated rice production. Evaluations of the greenhouse gas effects of SRI management with organic fertilization have found little or no increase in nitrous oxide emissions in the various SRI field trials [7].

SRI practices compared with conventional methods showed some notable benefits including response in root number (>30%), number of effective tillers in a hill (>25%), days to flowering (10 d earlier while counting days after seeding), and harvest index. In addition, SRI practices were effective in minimizing the incidence of rice leaf folder (1 larvae compared with 25 in conventional cultivation), shortening the rice crop cycle (by 8 d), and improving plant stand (10% lodging compared with 55% in conventional cultivation). However, grain yield was not significantly different between cultivation methods (6.3 t/hm^2 versus 6.7 t/hm^2 from conventional method). Except for harvest index (56% compared to 51% from inorganic management) and plant lodging percentage (9% compared to 56% in inorganic management), no significant effects were observed from the different management treatments [8].

Under SRI practices, the modified water saving techniques were able to increase rice yield by 0.49 t.ha^{-1} compared to traditional techniques. It is calculated that SRI can save a total of 2,193 m^3 of water per ha during the rice-growing season, a saving of 22%. Irrigation water was reduced by 1,933.5 m^3 ha^{-1} (by 23%). Water productivity was 1.12 kg of grain per m^3, with an increase of 0.30 kg per m^3 (36%), and irrigation water use efficiency was 1.34 kg of grain per m^3, an increase of 0.37 kg per m^3 [7].

The system of rice intensification is a rice production methodology that can be used by farmers to increase the water productivity in rice. Drying of rice paddies for between 4 and 12 days under SRI has positive impacts on rice yields. This results in water saving of between 27% and 42%. This saving has an implication on increasing area under rice irrigation [8].

8.2.3 MULCHING

Mulch is any type of material which is spread or laid over the surface of the soil to retain moisture in the soil, suppress weeds, and keep the soil cool. Mulch is usually but not exclusively organic in nature. It may be permanent like plastic sheet or temporary like straw mulch. Since agriculture is the main user of fresh water in the World, the use of water-saving irrigation methods can help to save water, which can be used in other sectors of the economy. In conventional rice cultivation, which is the most important irrigated crop, a significant amount of irrigation water is lost due to percolation and evaporation. So mulching can be treated as a better water saving technology along with weed management in rice.

A field experiment was conducted to study the effects of different mulches on growth and yield of tomato, weed growth, soil moisture and temperature. Polyethylene mulches were found superior to rice straw or sugarcane trash mulch in improving the growth and yield of tomato. Early flowering, greater number of fruits per plant and larger fruit size was observed with black and clear polyethylene mulch which resulted in 57.5 and 40.7% higher yield compared with the control (non-mulched). Black polyethylene mulch completely suppressed the weed growth, whereas clear polyethylene, rice straw and sugarcane trash mulches checked the weed growth to the extent of 70.2, 79.1 and 84.2%, respectively as compared to control. Higher soil temperature (2–3°C above the control) and soil moisture (43.7–62.5% higher than control) were also observed with polyethylene mulches. Natural mulching materials such as, rice straw or sugarcane trash also fetched an appreciable profit but black polyethylene has proved the most economical mulch [5, 6].

A field experiment was conducted to study water use efficiency and agronomic traits in rice cultivated in flooded soil and non-flooded soils with and without straw mulching. The total amount of water used by rice under flooded cultivation was 2.42 and 3.31 times as much as that by rice under the non-flooded cultivation with and without straw mulching, respectively. The average water seepage was 13560 m³/ha under the flooded cultivation, 4750 m³/ha under the non-flooded cultivation without straw mulching and 4680 m³/ha under non-flooded cultivation with straw mulching. Compared with the non-mulching treatment, straw mulching significantly increased leaf area per plant, main root length, gross root length and root dry weight per plant of rice. The highest grain yield under the straw mulching treatment (6747 kg/ha) was close to the rice cultivated in flooded soil (6 811.5 kg/ha) [18].

The effects of high density polyethylene (HDPE) film on increasing rice production, controlling weeds and residue amount of plastic were studied. The results indicated that the HDPE film mulching had significant effects on weed control, soil temperature, soil moisture, photosynthetic rate, seedling biomass and crop yield. Combined with economic effect, it showed that the HDPE film of 10 μm is the best option for rice production. Densities of total weeds were reduced significantly by application of mulches and dry weight of weeds was also significantly affected by the use of mulches. Different mulching techniques significantly improved the agronomic traits of aerobic rice. Plastic sheet mulching resulted in maximum paddy yield (4.18 t/ha) due to improvement in plant height (97.56 cm), number of panicles (25.73) and 1000-grain weight (18.43 g) [11].

Drip irrigation had a higher grain yield and harvest index, more effective tillers, more roots in topsoil, higher WUE, and greater economic benefit compared with the plastic mulching with furrow irrigation and non-mulching with furrow irrigation. Therefore, the drip irrigation could be considered a better water-saving technique in areas of arid and semiarid region.

Application of mulch increased the Fe-concentration and uptake in grain and straw in aerobic rice compared to no mulch aerobic rice treatment. Transplanted rice produced more yield compared to all aerobic rice treatments. But application of wheat straw/Sesbania mulch recorded significant higher yields compared to no mulch aerobic rice treatment. Growing of aerobic rice with Sesbania mulch and Fe fertilization produced higher grain and straw yield of aerobic rice with sufficient Fe nutrition [19].

According to a study conducted at University of Agriculture – Faisalabad – Pakistan, transplanted rice significantly lowered total weeds and weeds dry weight compared with direct seeded rice (no mulch) while plastic mulch among other mulch treatments had lowered total weeds and weeds dry weight. However, rice yield parameter such as plant height, fertile tillers, panicle length, and number of grains per panicle, test weight and total grain yield were significantly higher in Transplanted rice than direct seeded rice (no mulch) while plastic mulch had higher yield attributes than other mulch treatments. Transplanted rice and Direct seeded rice (sunflower mulch) showed maximum net returns and benefit cost ratio than other treatments [18].

Plot experiments and field investigations were conducted in 2011 and 2012 to investigate rice root spatial distribution at flowering and grain yield under plastic mulch with drip irrigation. The results showed that grain yield ranged from 3.35×10^3 kg ha^{-1} to 6.86×10^3 kg ha^{-1} under plastic mulch with drip irrigation, which was 19.3–60.31% lower than that under flooding irrigation [12].

8.3   MATERIALS AND METHODS

8.3.1   EXPERIMENTAL SITE

Field experiments at Swastik farm, Ranpur, District Nayagarh were conducted during Jan 2015 to May 2015. The site is located at latitude of 20°13′N and longitude of 85°1′E and an altitude of 20 m with respect to mean sea level. Before conducting experiments, the land was in fallow condition. The experiments on different cultivation practices with different irrigation levels were conducted in upland situation. At first the water requirement of hybrid rice crop was estimated under various cultural and irrigation practices. Then yield and yield attributes of rice crop under above cultural and irrigation practices were observed and recorded which will help us in developing crop-production models. The study comprised of estimating the crop water requirement of rice under various micro irrigation systems with different levels of irrigation along with the study of cost-effectiveness of different micro-irrigation systems in rice crop.

8.3.2   SOIL

The soil type of the experimental site is sandy loam and slightly acidic in nature which is taxonomically grouped under the order Alfisol. It is partly eroded due to high intensity of rainfall in the area. Geologically the soil is derived from laterite. The general slope of the land at the experimental site is 2%. The physical and chemical properties of soil of the experimental site were presented in Tables 8.2 and 8.3, respectively.

TABLE 8.2   Physical Properties of the Soil at the Experimental Site

Name of parameters	Measured value
Sand (%)	59.5
Silt (%)	26.2
Clay (%)	14.3
Bulk density (g cm^{-3})	1.55
Particle density (g cm^{-3})	2.64
Proctor moisture content (% by weight)	9.05

TABLE 8.3   Chemical Properties of the Soil at the Experimental Site

Parameter	Measured value	Remarks/comments
Available Nitrogen	142.60	Very Low
Available P	7.94	Low
Available Potassium	118.90	Low
Copper (ppm)	0.503	Adequate
EC (dS/m)	0.10	Good soil
Iron (ppm)	1.102	Adequate
Manganese (ppm)	2.732	Marginal
Organic Carbon (%)	0.60	Medium
pH	5.60	Medium Acidic
Zinc (ppm)	1.120	Marginal

8.3.3   CLIMATE

The climate of the study area is very pleasant and the weather is suitable for a wide range of crops available in Odisha. Generally the climate of the study area is humid and sub-tropical in nature. The average annual rainfall at the site is about 1250 mm. Eighty percent of the total rainfall occurs during the month of June to September in almost every year which is popularly known as monsoon rain fall. It experiences typical tropical weather conditions and succumbs to the heat and cold waves that sweep in from north India. The summer months from March to May are very hot and humid, and temperatures often rise above 40°C in the month of May. The South West monsoon lashes Odisha in the second week of June, bringing relief to the parched environment of the area. The study area receives maximum rainfall in the month of July and August whose average value is to the tune of 220 mm per month. The range of maximum and minimum atmospheric temperatures are from 34°C to 42°C and 24°C to 28°C in summer, 30°C to 36°C and 24°C to 27°C in monsoon and 22°C to 28°C and 9°C to 18°C in winter season, respectively. The range of maximum and minimum relative humidity vary from 77 to 92% and 29 to 53% in summer, 90 to 99% and 60 to 75% in monsoon, and from 80 to 95% and 30 to 55% in winter season, respectively.

8.3.4 WATER SOURCE

There are two bore wells and one farm pond in the experimental area. For this study, one bore well was used for irrigating rice crop at the experimental site. The chemical composition of runoff water and the bore well water at the experimental site are presented in Table 8.4.

8.3.5 CROP AND VARIETY SELECTION

The study area suffers from scarcity of water for irrigation purpose during the non-rainy seasons. The local farmers usually prefer to grow paddy in water scarce situation during *kharif* season as paddy crop is very much linked to their culture and tradition and they grow different vegetables during the *rabi* season by using certain alternate irrigation arrangements like continuous furrow, alternate furrow, drip and sprinkler irrigation systems, etc. Keeping the objectives of the research work in view, paddy crop was selected in *rabi* season with SRI method of cultivation at different irrigation levels to provide food security to each and every citizen of India in water scarce situation. A hybrid variety Arize6444 from Bayer seed Company was selected. It is a mid-duration hybrid, which gives higher yield, has good tolerance to water stress condition and provides good quality rice as certified by Ministry of Agriculture, Government of India, New Delhi.

8.3.6 TECHNICAL PROCEDURE

There were 32 subplots in total and area of each sub plot was 40 m². The experiment was designed with four treatments, four irrigation levels and

TABLE 8.4   Chemical Composition of Runoff Water and Bore Well Water at the Experimental Site

Parameter	Analysis result	Remarks	Analysis result	Remarks
	Runoff Water		**Water from bore well**	
Bi-Carbonate milli equiv. per liter	0.9	Less than 1.5 (Safe)	0.9	Less than 1.5 (Safe)
Calcium milli equiv. per liter	1.18	Less than 1.25 (Safe)	1.18	Less than 1.25 (Safe)
Carbonate milli equiv. per liter	1.48	Less than 1.5 (Safe)	1.48	Less than 1.5 (Safe)
Chloride milli equiv. per liter	0.25	Less than 2 (Safe)	0.25	Less than 2 (Safe)
EC (dS/m)	0.22	0.00 to 0.25 (Safe)	0.22	0.00 to 0.25
Magnesium milli equiv. per liter	2.28	Less than 5 (Safe)	2.28	Less than 5 (Safe)
pH	7.32	6.5 to 7.5 (alkaline)	7.32	6.5 to 7.5
Potassium milli equiv. per liter	—	—	—	
Residual Sodium Carbonate (RSC)	—	Less than 1.25 (Safe)	—	Less than 1.25 (Safe)
Sodium Absorption Ratio (SAR)	2.39	Less than 10 (Safe)	2.39	Less than 10 (Safe)
Sodium milli equiv. per liter	0.04	Less than 4 (Safe)	0.04	Less than 4 (Safe)
Sulphate milli equiv. per liter	—	Less than 2 (Safe)	—	Less than 2 (Safe)

two replications. Paddy seedlings of 15 days old were transplanted on raised beds of 15 cm height at the rate of single seedling per hill with a spacing of 20 x 20 cm^2.

Two lateral pipes of diameter 16 mm and length 10 m length each were laid on each raised bed at a spacing of 0.5 m. The lateral was inline lateral with inbuilt dripper of discharge 1.3 lph with a spacing of 0.3 m. The ground water was pumped with the help of a 2 HP submersible pump set and the water was supplied to the lateral line through main and submain line of the drip system. The quality of the irrigation water was good according to the water testing report at the site. The hybrid seed was procured directly from the Bayer seed Company, Bhubaneswar which is named as Arize 6444. The variety is drought and disease resistant. The experimental design was Randomized Block Design (RBD) as shown in Figure 8.1.

The nursery was prepared on 3rd January 2015 with a seed rate of 2.5 kg/acre. The date of transplanting of seedling was 18th January 2015. The row-to-row and plant-to-plant spacing was 20 cm. The duration of the crop was 126 days. In one of the four treatments, drip irrigation system in combination with black colored plastic mulching sheet of 50 micron thickness was used to reduce crop water demand significantly. Every day, the rice crop was irrigated after monitoring the status of soil moisture with the help of digital moisture meter. In the controlled plot, the rice crop was irrigated daily at 0% MAD, 10% MAD, 20% MAD and 30% MAD levels.

8.3.7   CULTURAL PRACTICES

8.3.7.1   Raising Nursery

8.3.7.1.1   Selection of Site

In SRI method, utmost care should be taken in the preparation of nursery bed. The nursery bed was prepared in the corner of the plot for quick and efficient transplanting to save time and transportation cost. Figure 8.2 shows view of uprooting of 15 days old seedlings from nursery. Hybrid rice of variety Arize 63444 was selected for study. The 14–15 days old seedlings (2–3 leaves stage) were transplanted.

8.3.7.1.2   Size of Bed

For one acre transplanting, the nursery beds were raised to a height of 10 cm with respect to ground level in 40 m² plot. For the experimental study, two raised beds each measuring 10 m² was prepared. Each bed had dimension of 5 m x 2 m. The height of the raised beds was restricted to 10 cm as the roots of 14–15 days old seedlings may grow up to a depth of 7.5 cm. To drain excess runoff water, resulting from heavy downpour, appropriate

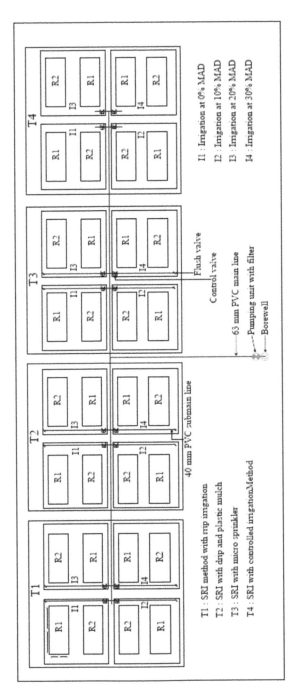

FIGURE 8.1   Layout plan of experimental setup (RBD).

FIGURE 8.2   Hybrid rice variety (Arize 6444); and uprooting of seedlings of 15 days old from the nursery bed (left).

channels were constructed on all sides of the nursery bed to safeguard the seedlings from submergence.

Nursery Bed Preparation: Nursery beds were prepared with application of farm yard manures (FYM) at the rate of 10t/ha and soil in four alternating layers. All these layers were mixed thoroughly as it will help in easy penetration of the roots of seedlings. Besides this, vermi-compost was uniformly spread over the two nursery beds to grow healthy seedlings.

8.3.7.1.3   Seed Rate

The 2.5 kg of seed is required to transplant in one acre of land. Therefore for the experimental study, 0.75 kg of seed was used for thin uniform spreading on these two raised beds. The sowing operation was performed in such a way that there was no crowding of seeds in any part of the nursery beds.

Seed Treatment: Healthy and pure seeds were soaked continuously for twelve hours in normal water. Then the water was drained off and the seeds were treated with Bavistin (2 g/kg of seed). Afterwards, the treated seeds were transferred to a water soaked gunny bag and kept for 24 hours. Finally the sprouted seeds were taken to the nursery bed for sowing.

8.3.7.1.4   Mulching and Watering

The nursery beds were covered with paddy straw to protect the sprouted seeds from direct sunlight and also to ensure protection from birds. Water was applied with rose cans twice daily, i.e., at morning at 6 AM and at evening 5 PM. Water was gently sprinkled so that the seeds should not come out while performing watering work. The paddy straw was removed from the nursery bed after two to three days of covering to ensure 100% germination.

8.3.7.2   Preparation of Main Field

The total experimental plot was plowed twice, i.e., first plowing by tractor mounted disc plow and second plowing by power tiller. Then soil was pulverized well by using rotavator. The land was uniformly leveled. Peripheral bunds were constructed around each sub plot of 40 m^2. There were 32 number of sub plots in total as per the experimental design. For first three treatments bed raiser was used for preparing raised beds. The dimension of each bed for transplanting was 10 m length, 1 m wide and 0.15 m height. Organic manures like farm yard manure or vermi-composts at the rate of 10 t/ha were applied 15 days before transplanting and mixed with the soil thoroughly during plowing operation. For the last treatment four number of sub plots were prepared by giving bunds around periphery up to a height of 10 cm and FYM were also added at the rate of 10 t/ha. Each sub plot was of 40 m^2 with dimension of 10.0 x 4.0 m^2.

8.3.7.3   Installation of Drip Irrigation System in Plots

8.3.7.3.1   Treatment 1

The drip irrigation system was installed before transplanting. The system constitutes of main line, sub main line, laterals, valves and accessories. The main line was 50 mm PVC pipe with pressure rating of 6 kg/cm^2 and the sub main line was 40 mm PVC pipe of 6 kg/cm^2 pressure rating. Design of the pipe line was based on the flow rate, length of the pipe line and the head loss in the pipe line. 16 mm inline laterals were used with dripper spacing of 30cm and dripper discharge of 1.3 lph. Two laterals per bed were laid out with a lateral spacing of 0.5 m.

8.3.7.3.2   Treatment 2

For the second treatment, drip lines were laid out as in the first treatment. Then, plastic mulching sheet of 50 micron thickness was covered on the raised bed. Holes are made in the mulching sheet at an interval of 20 cm. Figure 8.3 shows installation of drip irrigation in field (left) and layout of laterals on the raised bed system (right).

8.3.7.3.3   Treatment 3

The Micro Sprinkler system was installed before transplanting. The system was constituted of main line; sub main line, laterals, valves and accessories. The main line was 50 mm PVC pipe with pressure rating of 6 kg/cm^2 and the sub main line was 40 mm PVC pipe of 6 kg/cm^2 pressure rating. Design of the pipeline was based on the flow rate, length of the pipe line and the head loss in the pipe line. The 16 mm plain laterals were used with micro sprinklers at a spacing of 3 m and sprinkler discharge of 44 lph. Laterals

FIGURE 8.3   Installation of drip irrigation (left) and laying of laterals on the raised bed (right).

were placed at a spacing of 3 m. In each plot, a digital moisture meter was installed to know moisture status of each sub plot under treatment-3 and accordingly each irrigation was provided for an average period of 10 min to maintain moisture level at a particular MAD level. Figure 8.4 shows the view of raised soil beds covered with mulching sheet having holes so that seedlings can come through them; and testing of micro sprinkler (right).

8.3.7.4 Irrigation Systems for Controlled Plot

The 50 mm PVC pipe with pressure rating of 6 kg/cm^2 was laid in the four subplots with individual control valve. The discharge capacity of pipe was 1.7 lps.

8.3.7.5 Design of Head Unit

A submersible pump of 2 HP was used to supply water to the drip network in this research work. The source of power for pumping was electricity. Pump capacity was determined on basis of crop water requirement, system efficiency and area to be irrigated in a given time. Similarly, discharge was determined based on the desired operating pressure, functional head loss

FIGURE 8.4   Raised beds covered with mulching sheets having punched holes (left); and testing of micro sprinkler (right).

and change of elevation within the field. The filter with capacity 25 m³/h was used in the head unit. As the water quality is good, no other filters were required for the research work.

8.3.7.6   Transplanting

Light irrigation was provided before transplanting so that main paddy field should be in wet condition but no standing water in the field. Young rice seedlings of 15 days old (2–3 leaves stage) were transplanted in various treatments (Figure 8.5). The seedlings with 2–3 leaves stage have great potential for profuse tillering and root development. It results in achieving maximum yield potential of hybrid rice varieties. Before transplanting, a rope was used as marker to lay out the plot into square grid of 20 cm x 20 cm size, i.e., row-to-row and plant-to-plant spacing.

8.3.8   IRRIGATION SCHEDULING

Irrigation scheduling was done by taking the moisture reading on daily basis. In the study, there were four irrigation regimes. Valve operating time was maintained accordingly to the required moisture level in the soil. The different irrigation levels were:
- Irrigation at 0% management allowable deficit (MAD) level of available soil moisture (ASM), i.e., at field capacity (I_1)
- Irrigation at 10% MAD Level (I_2)
- Irrigation at 20% MAD Level (I_3)
- Irrigation at 30% MAD Level (I_4)

In each case, irrigation was stopped when the soil moisture was at field capacity. A digital soil moisture meter was used to monitor the soil moisture content on daily basis (Figure 8.6). Moisture contents were recorded in the surface layer only to a depth of 0 to 30 cm.

8.3.9   WEED MANAGEMENT

As there was no standing water in SRI method, weeds were more in the standing water paddy crop. Therefore in this study, chemical weedicide was

Treatment-1   Treatment-2

Treatment-3   Treatment-4

FIGURE 8.5   Transplanting of 15 days-old rice seedling in each of the four treatments

FIGURE 8.6   Moisture meter in the experimental plot.

sprayed on the raised bed three days after transplanting. As the plot size was small, manual weeding was done twice on 10th and 20th day after transplanting.

8.3.10. FERTIGATION SCHEDULING

Fertilizers in the experiment were 180:80:80 of N: P: K kg/per acre after the soil testing. Fertigation dose was maintained constant in all treatments. Urea, Single Super Phosphate (SSP) and Mureate of Potash (MOP) were used as source of N, P and K, respectively. Five doses of fertilizers were applied throughout the growing season (Table 8.5).

8.3.11 HARVESTING

The fully matured rice crop was harvested manually on 126th day and was stacked separately according to each sub plot. Then the harvested crop was dried in direct sun light for three days for further processing like threshing and cleaning. Figure 8.7 shows the views of matured rice crop and harvested crop in the field.

8.3.12 EXPERIMENTAL DATA

8.3.12.1 Growth Parameters

Five hills per sub plot were selected randomly and tagged for recording growth parameters at different growth stages after transplanting up to har-

TABLE 8.5   Fertilization Program the Experimental Site

Date of fertilizer application	Urea (kg)	MOP (kg)	SSP (kg)
17th Jan 2015	0	50	0
28th Jan 2015	6.5	–	0
26th Feb 2015	20	–	2
15th Feb 2015	6.5	–	2
10th Apr 2015	6.5	–	5

FIGURE 8.7 View of matured rice crop (left) and harvested crop in the field (right).

vesting. The plant height from five hills were measured from the ground level to tip of the topmost leaf at different growth at 30, 50, 90 and 120 DAT (days after transplanting). Then the average plant height per hill was calculated (Table 8.6). The number of effective tillers in five hills was counted and the average was worked out (Table 8.7). The root length from five hills was measured and the average was worked out (Table 8.8).

8.3.12.2 Yield and Yield Attributes

Yield of 32 sub plots, panicle length, number of grains per panicle, test weight of 1000 grains, and mass of dry straw (biomass) were recorded for the analysis of data. In each sub plot, five hills were randomly selected. From each hill, five numbers of panicles from effective tillers were selected and length of each panicle was measured and the average panicle length per sub plot was found out.

The numbers of panicles from 25 hills (5 hills/subplot) were counted. The mean values were worked out (Table 8.9). Panicle length measurements (cm) were recorded from base of the panicle to the tip of the panicle. The mean value was calculated (Table 8.10).

One thousand grains were counted from randomly selected five hills per sub plot and their test weight was recorded (Table 8.11). The grains were separated by threshing separately from each sub plot of 40 m² area and were dried under direct sunlight for three days. The moisture content of grains was measured using moisture meter and was found to be

TABLE 8.6 Plant Height (cm) at Different Critical Stages of Rice as Influenced by Different Methods and Levels of Irrigation

Treat-ment	Irrigation level	Active tillering	Panicle initiation	Flowering	Maturity	Average plant height at maturity
				cm		
T1	I1	37.40	48.85	74.90	96.10	100.70
	I2	40.25	51.80	79.10	108.40	
	I3	35.10	45.50	71.10	99.35	
	I4	35.10	45.45	67.55	98.95	
T2	I1	38.05	49.35	79.90	100.85	99.65
	I2	41.10	53.20	80.45	101.95	
	I3	34.90	45.55	70.55	98.05	
	I4	34.95	45.40	68.95	97.75	
T3	I1	40.05	51.90	70.60	98.73	100.94
	I2	42.25	55.33	76.90	107.85	
	I3	39.90	52.10	72.15	100.65	
	I4	37.40	48.83	69.15	96.53	
T4	I1	29.90	39.15	59.70	102.55	102.72
	I2	40.10	51.90	79.10	110.69	
	I3	37.90	49.45	74.10	99.20	
	I4	35.25	45.55	71.35	98.45	

10%. Later, winnowing and cleaning operations were carried out to get cleaned paddy. Then the weight of cleaned paddy for each sub plot was measured and recorded for analysis and the recorded data are presented in the Table 8.12. Data for dry straw yield as influenced by different methods and levels of irrigation are presented in Table 8.13. Water productivity for different methods and levels of irrigation are presented in Table 8.14. The statistical analysis of yield and yield attributes were made at different levels of irrigation for each treatment using SPSS 16.0 software.

TABLE 8.7 Number of Effective Tillers/Plant as Influenced by Different Methods and Levels of Irrigation

Treat-ment	Irrigation level	30 days after planting	40 days after planting	50 days after planting	Average number of tiller at 50 DAP
T1	I1	14.5	23.5	38.5	37
	I2	15.5	26.5	40.5	
	I3	15.5	23.5	34.5	
	I4	14	22	34.5	
T2	I1	14.5	22.5	36	35
	I2	14.5	24.5	38.5	
	I3	15	21.5	34.5	
	I4	12.5	20.5	31	
T3	I1	13	20.5	40	39
	I2	14.5	23.5	40.5	
	I3	14.5	21.5	39.5	
	I4	13	20	36	
T4	I1	12.5	17.5	25	24
	I2	13.5	19.5	28.5	
	I3	11.5	16.5	23.5	
	I4	12	15	19	

8.4 RESULTS AND DISCUSSION

8.4.1 ESTIMATION OF WATER REQUIREMENT OF RICE

8.4.1.1 Irrigation Requirement

Irrigation requirements of the rice crop for all the treatments during the study period are given in Table 8.15. Irrigation requirement for SRI method of rice cultivation with controlled irrigation was found to be higher than all other treatments. The irrigation requirement in the controlled plot was found to be in the range of 702.60 mm (at 30% MAD level) to 1056.56 mm (at 0% MAD level). In case of SRI with drip and plastic mulch (T2), the irrigation water requirement was found to be the lowest at 30% MAD level (417.56 mm) whereas, the irrigation water requirement in case of

TABLE 8.8   Root Length (cm) at Different Critical Stages of Rice as Influenced by Different Methods and Levels of Irrigation

Treatment	Irrigation level	Active tillering	Panicle initiation	Flowering	Maturity	Average root length at maturity (cm)
T1	I1	18.50	18.50	21.50	27.28	26.53
	I2	15.80	19.60	22.55	26.85	
	I3	14.65	17.75	19.10	26.60	
	I4	13.10	16.00	18.10	25.40	
T2	I1	18.05	20.10	22.45	29.05	26.78
	I2	17.70	21.10	23.10	28.45	
	I3	15.90	18.65	21.90	26.30	
	I4	15.35	17.10	18.90	23.30	
T3	I1	15.40	20.45	22.25	26.85	25.33
	I2	15.90	18.85	22.20	27.55	
	I3	14.55	18.50	21.20	25.25	
	I4	13.10	16.05	20.10	21.65	
T4	I1	18.05	21.45	22.90	27.60	26.88
	I2	18.50	20.00	22.45	27.30	
	I3	16.95	19.50	21.95	26.55	
	I4	15.35	18.10	21.70	26.05	

SRI with controlled irrigation at field capacity level (T4) was found to be the highest.

The irrigation requirements in drip plots (T1) at 0, 10, 20, 30% MAD levels were observed to be 674.96, 606.32, 551.41 and 472.47 mm, respectively. Irrigation requirement at 0% MAD level was observed to be higher than all other irrigation levels because more irrigation water was required by the plant at this level. Total irrigation requirement of rice crop at 0% MAD level was found to be 56.53% more in case of SRI with controlled irrigation than that in case of SRI with drip irrigation at 0% MAD level. Similarly total irrigation requirement of rice in SRI with controlled irrigation at 0% MAD level was found to be 66.7% and 31.52% more as compared to SRI with drip and plastic mulch and SRI with micro sprinkler, respectively (Table 8.15).

TABLE 8.9   Number of Panicles per Hill as Influenced by Different Methods and Levels of Irrigation

Treatment	Irrigation level	Number of panicles per hill	Average number of panicles
T1	I1	33	33.00
	I2	37	
	I3	31	
	I4	32	
T2	I1	29	29.03
	I2	32	
	I3	29	
	I4	27	
T3	I1	32	31.00
	I2	33	
	I3	31	
	I4	29	
T4	I1	20	18.09
	I2	21	
	I3	18	
	I4	14	

8.4.1.2   Crop Water Requirement

The water requirement of the rice crop was computed as the sum of the irrigation water applied, effective rainfall, contribution from ground water through capillary rise and contribution from soil moisture (Water Balance Method). The effective root zone depth of rice was taken as 30 cm because there is maximum concentration of roots within this depth. The net irrigation requirements of the crop are calculated using the field water balance as below:

$$I_n = ET_{crop} - (P - R + Ge + W_b)$$ (1)

where, I_n is the net irrigation requirement of the crop; ET_{crop} is the crop evapotranspiration; P is precipitation; R is surface runoff; Ge is ground water contribution; and W_b is the soil moisture contribution.

In the above equation, groundwater contribution to crop root zone was neglected because the ground water table was at more than 10 m below the

TABLE 8.10   Panicle Length (cm) as Influenced by Different Methods and Levels of Irrigation

Treatment	Irrigation level	Panicle length, cm	Average panicle length, cm
T1	I1	29.75	28.63
	I2	30.51	
	I3	27.70	
	I4	26.55	
T2	I1	28.85	28.48
	I2	29.13	
	I3	28.68	
	I4	27.25	
T3	I1	28.20	26.80
	I2	27.50	
	I3	27.45	
	I4	24.05	
T4	I1	24.65	24.33
	I2	26.60	
	I3	23.35	
	I4	22.70	

effective root zone of the crop. The component $(P–R)$ is termed as effective rainfall. In the experimental site, total seasonal rainfall during the crop growth period was only 64 mm comprising of 8 rainfall events. Taking the potential evapotranspiration 8 mm per day during the cropping season, it was found that the effective rainfall was 64 mm. Water requirement of the crop for all the treatments are given in Table 8.16.

The crop water requirement of the SRI rice at 0% MAD level (controlled irrigation) was found to be the highest compared to other three micro irrigation treatments. The crop water requirement for SRI rice with controlled irrigation was 1120.00, 988.00, 916.00 and 766.00 mm at 0, 10, 20 and 30% MAD level, respectively. Water requirement for SRI with micro sprinkler was 907.84, 809.39, 756.65 and 641.80 mm at 0, 10, 20 and 30% MAD level, respectively. Crop water requirement for SRI with drip and plastic mulch was minimum because of less evapotranspiration and it was 670.32, 604.68, 563.93 and 481.56 mm at 0, 10, 20 and 30% MAD level, respectively. However, SRI rice with drip required more

TABLE 8.11 Test Weight (g/1000 grains) as Influenced by Different Methods and Levels of Irrigation

Treatment	Irrigation level	Test weight (g)	Average test weight (g)
T1	I1	37.65	36.38
	I2	38.30	
	I3	35.65	
	I4	33.90	
T2	I1	38.80	38.95
	I2	40.30	
	I3	39.30	
	I4	37.40	
T3	I1	38.20	37.98
	I2	39.20	
	I3	37.50	
	I4	37.00	
T4	I1	28.55	27.45
	I2	29.00	
	I3	26.65	
	I4	25.60	

water in comparison to SRI with drip and plastic mulch and it was 738.96, 670.32, 615.41 and 536.47 mm at 0, 10, 20 and 30% MAD level, respectively. There was a significant saving of 42% water in SRI with drip and plastic mulch and 34 % water in SRI with drip irrigation at 10% MAD level as compared to SRI with controlled irrigation.

8.4.2 COMPARISON OF YIELD AND YIELD ATTRIBUTES OF RICE UNDER VARIOUS IRRIGATION METHODS AT DIFFERENT MAD LEVEL

8.4.2.1 Growth Parameters of Rice

8.4.2.1.1 Plant Height

The data pertaining to plant height at different growth stages of the crop including maturity stage are presented in Table 8.17 and in Figure 8.8.

TABLE 8.12 Rice Yield (100 kg/ha) as Influenced by Different Methods and Levels of Irrigation

Treatment	Irrigation level	Yield, q/ha	Average yield, q/ha.
		One q/ha = 100 kg/ha	
T1	I1	82.58	81.25
	I2	83.41	
	I3	80.38	
	I4	78.63	
T2	I1	84.64	84.40
	I2	86.15	
	I3	84.25	
	I4	82.55	
T3	I1	74.65	76.80
	I2	80.60	
	I3	78.75	
	I4	73.20	
T4	I1	47.60	48.90
	I2	52.60	
	I3	49.70	
	I4	45.70	

The rice crop grew taller with the advancement in age and the increase in height was more rapid up to the flowering stage which occured at 90 days after planting (DAP). Average plant height at maturity was 102.72 cm in SRI method of cultivation with controlled irrigation while the average plant height was 99.65 cm in SRI with drip and plastic mulch. The reason of getting more height in SRI method of cultivation with controlled irrigation may be due to intercultural operations (loosening of soil and weeding), which was not feasible in case of SRI with drip and plastic mulch.

SRI with controlled irrigation at 10% MAD level recorded the highest average plant height of 40.0, 52.0, 79.2 and 110.88 cm at active tillering, panicle initiation, flowering and maturity stage, respectively. It was found to be superior over all other treatments. SRI with micro sprinkler recorded the second highest average plant height and it was followed by SRI with

TABLE 8.13 Dry Straw Yield (100 kg/ha) as Influenced by Different Methods and Levels of Irrigation

Treatment	Irrigation level	Dry straw yield, q/ha.	Average dry straw yield, q/ha.
			One q/ha = 100 kg/ha
T1	I1	33.82	36.56
	I2	41.73	
	I3	35.32	
	I4	35.38	
T2	I1	34.76	37.99
	I2	43.58	
	I3	37.03	
	I4	36.60	
T3	I1	30.99	34.61
	I2	40.30	
	I3	34.48	
	I4	32.67	
T4	I1	25.80	25.65
	I2	26.40	
	I3	25.75	
	I4	24.65	

drip irrigation and SRI with drip and plastic mulch, respectively. Irrigation at 0, 10, 20 and 30% MAD level in case of SRI with drip and plastic mulch revealed that the average plant heights were the lowest in succession in comparison to other three treatments.

From descriptive statistical analysis it was observed that the average plant height at active tillering stage for all irrigation level under treatment I was 36.96 cm with a standard deviation of 2.27 cm. Similarly, the average plant height at active tillering stage for all irrigation levels under treatment T2, T3 and T4 were 37.25 cm, 39.90 cm and 35.78 cm with standard deviation of 2.74 cm, 1.84 cm and 4.07 cm, respectively.

From statistical analysis it was evident that the average plant height for all irrigation levels under treatments T1, T2, T3 and T4 was 36.96, 37.25, 39.80 and 35.79 cm at standard deviation of 2.27, 2.74, 1.84 and 4.07 cm

TABLE 8.14   Water Productivity as Influenced by Different Methods and Levels of Irrigation

Treatment	Irrigation level	Grain yield, qt/ha.	Water used, m³	Water productivity, kg/m³	Average water productivity, kg/m³
T1	I1	83.58	10389.60	0.80	0.87
	I2	84.85	9703.20	0.87	
	I3	80.68	9154.08	0.88	
	I4	75.88	8364.72	0.91	
T2	I1	84.64	9703.20	0.87	0.96
	I2	86.15	9016.80	0.96	
	I3	84.25	8639.28	0.98	
	I4	82.55	7815.60	1.06	
T3	I1	74.65	12078.40	0.62	0.72
	I2	80.60	11093.92	0.73	
	I3	78.75	10566.52	0.75	
	I4	73.20	9417.96	0.78	
T4	I1	47.60	14200.60	0.34	0.40
	I2	52.60	12880.45	0.41	
	I3	49.70	12160.85	0.41	
	I4	45.70	10660.48	0.43	

TABLE 8.15   Irrigation Requirement of Rice Crop (mm) under Different Irrigation Practices with Different MAD Level

Treatment	MAD level			
	I1	I2	I3	I4
	Irrigation requirement of rice crop (mm)			
SRI with drip irrigation	674.96	606.32	551.41	472.47
SRI with drip and plastic mulch	606.32	537.68	499.93	417.56
SRI with micro sprinkler	843.84	745.39	692.65	577.80
SRI with controlled irrigation	1056.56	924.45	852.40	702.60

TABLE 8.16 Water Requirement of Rice Crop under Different Irrigation Practices with Different MAD Level

Treat- ment	Irriga- tion level	Irrigation requirement (mm)	Effective rainfall	Soil moisture contribution	Deep perco- lation	Water require- ment (mm)
T1	I1	674.96	64	0	0	738.96
	I2	606.32	64	0	0	670.32
	I3	551.41	64	0	0	615.41
	I4	472.47	64	0	0	536.47
T2	I1	606.32	64	0	0	670.32
	I2	537.68	64	0	0	601.68
	I3	499.93	64	0	0	563.93
	I4	417.56	64	0	0	481.56
T3	I1	843.84	64	0	0	907.84
	I2	745.39	64	0	0	809.39
	I3	692.65	64	0	0	756.65
	I4	577.80	64	0	0	641.80
T4	I1	1056.00	64	0	0	1120.00
	I2	924.00	64	0	0	988.00
	I3	852.00	64	0	0	916.00
	I4	702.00	64	0	0	766.00

for tillering; 73.16, 74.96, 72.20 and 71.06 cm at standard deviation of 4.60, 5.60, 3.12 and 7.62 cm for panicle initiation; and 100.70, 99.65, 100.94 and 102.72 cm at standard deviation of 4.94, 1.92, 4.54 and 5.18 cm for maturity stage, respectively. In most of the cases the average plant height is less in SRI with controlled irrigation (T4). The reason may be due to leaching of water and nutrients from the effective root zone of the crop. From analysis of variance it was found that the variation between treatments and variation between irrigation levels with regard to plant height at all stages (tillering, panicle initiation, flowering and maturity) was significant.

8.4.2.1.2 Number of effective tillers per hill

Data pertaining to the number of effective tillers per hill are given in Table 8.18 and in Figure 8.8. The rice crop showed that there was progressive

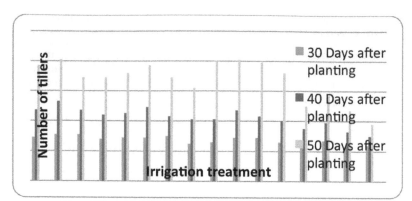

Top: Plant height at different growth stages of rice

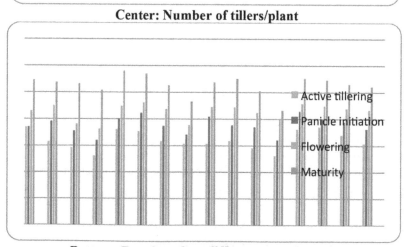

Center: Number of tillers/plant

Bottom: Root length at different growth stages

FIGURE 8.8 Growth Parameters at different growth stages of rice as influenced by different methods and levels of irrigation: Plant Height (top, cm); No. of tillers (center); and root length (bottom, cm).

TABLE 8.17 Average Plant Height (cm) at Different Growth Stages under Each Treatment Irrespective of Level of Irrigation

Growth stages	Treatment	Mean	Std. Deviation
Active tillering stage	I	36.97	2.26964
	II	37.25	2.74122
	III	39.9	1.84236
	IV	35.79	4.07446
Panicle initiation stage	I	47.90	2.82476
	II	48.38	3.42626
	III	52.04	2.46007
	IV	46.51	5.14988
Flowering stage	I	73.16	4.59998
	II	74.96	5.60941
	III	72.20	3.11907
	IV	71.06	7.61764
Maturity	I	100.7	4.93848
	II	99.65	1.92354
	III	100.9	4.54401
	IV	102.72	5.18939

increase in number of effective tillers up to 50 DAP. Comparatively more number of effective tillers was noticed in SRI method of cultivation with micro sprinkler (T3) than the other treatments. Average number of effective tiller was 40 in case of SRI method of cultivation with micro sprinkler while in case of SRI with controlled irrigation (T4), it was 24 cm.

SRI with micro sprinkler at 10% MAD level recorded the maximum number of effective tillers per hill at 50 DAP, which was found to be superior over other treatments followed by SRI with drip, SRI with drip and plastic mulch and SRI with controlled irrigation. The possible causes may be due to availability of proper quantities of water, nutrient and air at the root zone of the crop. Another reason may be due to creation of conducive environment above and below the soil which controls soil and air temperature in a better way along with humidity of air.

It was observed from the statistical analysis that the average number of active tillers at 30 DAP for all irrigation levels under treatment I, II, III

TABLE 8.18　Average Number of Effective Tillers PER Hill at Different Growth Stages in each Treatment Irrespective of Level of Irrigation: DAP = days after planting

Growth stages	Treatment	Mean	Std. Deviation
Number of effective tillers/hill 30 DAP	I	14.8750	.99103
	II	14.1250	1.24642
	III	13.7500	.88641
	IV	12.3750	.91613
Number of effective tillers/hill 40 DAP	I	23.8750	1.80772
	II	22.2500	1.66905
	III	21.3750	1.50594
	IV	17.1250	1.80772
Number of effective tillers/hill 50 DAP	I	37.0000	2.82843
	II	35.0000	2.97610
	III	39.0000	2.00000
	IV	24.0000	3.70328

and IV was 15, 14, 14 and 12 with a standard deviation of 0.99, 1.25, 0.89 and 0.92, respectively. Similarly, the average number of active tillers for all irrigation levels under treatment I, II, III and IV were found to be (24, 22, 21 and 17) and (37, 35, 39 and 24) with a standard deviation of (1.80, 1.67, 1.50 and 1.81) and (2.83, 2.98, 2.0 and 3.70), at 40 DAP and 50 DAP, respectively. Table 8.8 shows that the variation between treatments and variation between irrigation levels with regard to number of effective tillers at 30, 40 and 50 DAP were found to be significant.

8.4.2.1.3. Root Length at Critical Stages

The data pertaining to the root length at different critical stages of the crop are presented in Table 8.19 and Figure 8.8. The root length showed its progressive increase in SRI method of cultivation with controlled irrigation as compared to other irrigation practices (treatments). Comparatively higher root length was found in SRI method of cultivation with controlled irrigation as compared to all other methods of irrigation at all MAD levels. SRI method of cultivation with controlled irrigation (T4) showed the highest average root length of 26.88 cm while in case of SRI with micro sprinkler (T3) showed the lowest average root length of 25.33 cm.

TABLE 8.19 Average Root Length (cm) at Different Growth Stages in each Treatment Irrespective of Level of Irrigation

Growth stages	Treatment	Mean	Std. Deviation
Average root length at active tillering stage	I	15.51	2.13102
	II	16.75	1.23520
	III	14.74	1.13884
	IV	17.21	1.29883
Average root length at panicle initiation stage	I	17.96	1.43620
	II	19.24	1.61770
	III	18.46	1.68602
	IV	19.76	1.28167
Average root length at flowering stage	I	20.31	1.93127
	II	21.59	1.72249
	III	21.44	1.76023
	IV	22.75	1.03363
Average root length at maturity	I	26.83	1.70226
	II	26.78	1.20090
	III	25.33	2.40876
	IV	26.88	2.43882

SRI with controlled irrigation at 0, 10, 20 and 30% MAD levels recorded the maximum root length of 23.22, 29.00, 22.00 and 21.80 cm, respectively, which was observed to be superior over all other treatments at respective MAD levels (Table 8.19). From statistical analysis, it was observed that the average root length under SRI with controlled irrigation (T4) was found to be higher in comparison to other treatments at different critical stages of the crop growth.

The reason of getting higher average root length may be due to movement of irrigation water to deeper layers (below the root zone) as compared to other treatments. In other treatments, water is available at shallow depth due to controlled application of water through emitters which restricts root growth vertically and encourages it laterally.

The percentage of root growth was more pronounced from flowering to maturity stage, i.e., to the tune of 39.5% as compared to other critical stages irrespective of any treatments. From analysis of variance, it was found that the variation between treatments and variation between irrigation levels with regard to average root length at all growth stages (tillering, panicle initiation, flowering and maturity) was found to be significant (Table 8.19).

8.4.2.2   Yield and Yield Components

8.4.2.2.1   Panicle Length

The data pertaining to average panicle length (cm) of rice crop are presented in Table 8.20 and in Figure 8.9a. It was noticed that there was not much variation in panicle length among different irrigation levels within a particular method of irrigation. But there was clear-cut variation among different methods of irrigation. The reason may be due to the availability of nutrients in different methods of irrigation. The maximum average panicle length was recorded in SRI method of cultivation with drip (28.63 cm) and the lowest was recorded in SRI with controlled irrigation (24.33 cm).

It was found that SRI with drip irrigation at 10% MAD level recorded the maximum panicle length of 30.50 cm, which was comparatively superior over other treatments followed by SRI with drip and plastic mulch, SRI with micro sprinkler and SRI with controlled irrigation. SRI with controlled irrigation at 30% MAD level recorded the lowest panicle length (23.00 cm). It was found from the statistical analysis that the average length of panicles under treatments T1 to T4 was 28.63, 28.45, 26.80 and 24.33 with standard deviation of 1.69, 0.78, 1.74 and 1.61, respectively. Further it was noticed that the average length of panicle in case of SRI with drip was found to be more due to better root development as compared to other treatments (Table 8.20). From Table 4.6 it was found that the variation between treatments and variation between irrigation levels with regard to length of panicles at maturity stage was found to be significant.

8.4.2.2.2   Number of Panicles

Data pertaining to number of panicles are presented in Table 8.20 and in Figure 8.9b. There is not much variation in the number of panicles under

TABLE 8.20 Average Values of Yield and Yield attributes under each Treatment Irrespective of Level of Irrigation

Treatment	Mean	Std. Deviation
Average length of panicle (cm)		
I	28.6275	1.68906
II	28.4750	0.77736
III	26.8000	1.73864
IV	24.3250	1.60513
Average number of panicles/hill		
I	33.1250	2.16712
II	29.0000	2.13809
III	31.0000	1.41421
IV	18.1250	3.13676
Average test weight, g per 1000 grains		
I	36.3750	1.86452
II	38.9500	1.13389
III	37.9750	.89881
IV	27.4500	1.50238
Rice yield, q/ha or 100 kg/ha		
I	81.2450	2.00519
II	84.3975	1.37469
III	76.8000	3.20491
IV	48.9000	2.74278
Dry straw yield, q/ha or 100 kg/ha		
I	36.5612	3.26254
II	37.9888	3.56731
III	34.6088	3.75754
IV	25.6500	.68452

Note: In this chapter, one qt. (quintal) = 100 kg.

different methods of irrigation. But the variation was noticed under different levels of irrigation. The average number of panicles in SRI method of cultivation with drip was found to be the highest, i.e., 33.0 whereas, SRI with controlled irrigation recorded the lowest number of panicles with an average of 18.0. It was observed that SRI with drip irrigation at 10% MAD

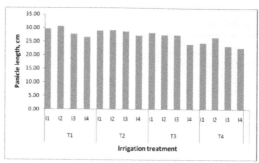

Fig. (a) Panicle length as influenced by different methods and levels of irrigation

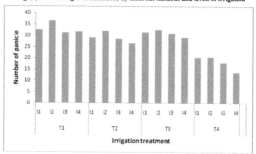

Fig. (b) Number of panicles/hill as influenced by different methods and levels of irrigation

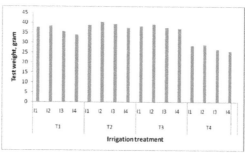

Fig. (c) Test weight as influenced by different methods and levels of irrigation

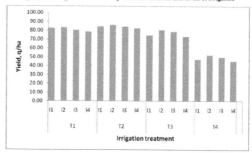

Fig. (d) Rice yield as influenced by different methods and levels of irrigation

FIGURE 8.9   Rice yield (100 kg/ha) and its attributes as influenced by different methods and levels of irrigation.

level recorded the maximum number of panicles of 36.0. It was comparatively superior over other treatments followed by SRI with drip irrigation and plastic mulch, SRI with micro sprinkler and SRI with controlled irrigation. SRI with controlled irrigation at 30% MAD level recorded the lowest number of panicles of 14.0.

Regarding average number of panicles per hill in Table 8.20, it was found that average numbers of panicles were maximum in SRI with drip (T1) followed by SRI with drip and plastic mulch (T2). The number of panicles per hill was minimum in case of SRI with controlled irrigation. The reason may be due to loss of nutrients and water below the root zone of the crop resulted from deep percolation. The variation between treatments and between irrigation levels with regard to number of panicles at maturity was significant (Table 4.7).

8.4.2.2.3 Test Weight

The data pertaining to test weight of 1000 grains are presented in Table 8.20 and in Figure 8.9c. There was much variation under different methods of irrigation as well as different levels of irrigation. The average 1000-grain weight was observed to be the highest in SRI method with drip, i.e., 36.38 g as compared to SRI method with controlled irrigation (27.45 g). It was found that SRI with drip irrigation at 10% MAD level recorded the maximum test weight of 38.40 g, which was comparatively superior over other treatments followed by SRI with drip and plastic mulch, SRI with micro sprinkler and SRI with controlled irrigation, respectively. SRI with controlled irrigation at 30% MAD level recorded the lowest test weight (25.40 g).

From Table 8.20, it is observed that the average test weight of 1000 grains in case of SRI with drip and mulch (T2) was higher as compared to other treatments due to better availability of nutrients and water as evidenced from the least weeding problem.

8.4.2.2.4 Yield

The data pertaining to yield of hybrid rice crop are presented in Table 8.20 and in Figure 8.9d. There was significant variation among different methods of irrigation as well as levels of irrigation. In this chapter, one quintal (q) is equal to 100 kg.

The average yield was highest in SRI with drip irrigation and plastic mulch (84 q/ha) as compared to SRI with controlled irrigation (48 q/ha). It was observed that SRI with drip irrigation and plastic mulch at 10% MAD level recorded the maximum yield of 86 q/ha, which was comparatively superior over other treatments followed by SRI with drip irrigation, SRI with micro sprinkler and SRI with controlled irrigation. The reason may be due to absence of weeds and higher nutrient use efficiency. SRI with controlled irrigation at 30% MAD level recorded the lowest yield (44 q/ha.)

It was revealed that higher yields were obtained in case of SRI with drip and plastic mulch (T2), i.e., to the tune of 84.4 q/ha with least standard deviation of 1.37 as compared to the other treatments followed by SRI with drip (81.25 q/ha), SRI with micro sprinkler (76.8 q/ha) and SRI with controlled irrigation (48.9 q/ha). Multiple regression analysis indicated that yield is a function of root length at active tillering stage, root length at flowering stage, root length at maturity stage, length of panicle and test weight of 1000 grains as these significantly contribute towards rice yield. The other growth parameter like plant height at all growth stages, number of tillers at 30, 40 and 50 DAP, root length at panicle initiation stage, number of panicles per hill and dry straw yield did not significantly contribute to the yield as evidenced from analysis. The model explains 99.8% variation ($R^2 = 0.998$) in the data with a standard error of estimate of 0.921.

The mathematical model for the yield is stated below:

$$Y = (-0.555) \times X_1 + (-1.268) \times X_2 + 0.969 \times X_3 \\ + 0.825 \times X_4 + 1.713 \times X_5 - 10.272 \qquad (2)$$

where, Y = Yield of the crop, q/ha; X_1 = Root length at active tillering stage, cm; X_2 = Root length at flowering stage, cm; X_3 = Root length at maturity stage, cm; X_4 = Panicle length, cm; and X_5 = Test weight, g.

8.4.2.2.5   Dry Straw Yield

The data pertaining to dry straw yield was presented in Table 8.20. Much variation was noticed under different methods of irrigation as well as levels of irrigation with respect to straw yield. The average straw yield was recorded to be the highest in SRI with drip and plastic mulch (41.91 q/ha) as compared to SRI with controlled irrigation (26.48 q/ha). The reason may be due to better growth of plants and uniformity in plant height.

It was found that SRI with drip and plastic mulch at 10% MAD level recorded the maximum yield of 43.32 q/ha, which was comparatively more over other treatments followed by SRI with drip irrigation, SRI with micro sprinkler and SRI with controlled irrigation. SRI with controlled irrigation at 30% MAD level recorded the lowest yield (25.54 q/ha.)

The Table 8.20 shows that dry biomass yield in case of SRI with drip and plastic mulch was to the tune of 38 q/ha followed by SRI with drip (36.56 q/ha), SRI with micro sprinkler (34.61 q/ha) and SRI with controlled irrigation (25.65 q/ha).

8.4.2.3   Interaction Effect of Treatments and Irrigation Levels on Yield and Yield Attributes

The interaction effect due to treatment and irrigation levels was found to be significant in case of plant height at all growth stages at 1% level of significance. The number of tillers at 30 and 40 days after planting was non-significant. But the number of tiller at 50 days after planting was significant at 5% confidence level. The root length at panicle initiation, flowering and maturity stage was found to be significant at 1% level of significance. But root length at tillering stage was found to be non-significant as evidenced from Table 8.21. The other parameters like panicle length, number of panicles, test weight, yield and dry biomass were found to be significant at 1% level of significance.

8.4.2.4   Grand Mean of Yield and Yield Attributes

The grand mean was calculated taking all biometric parameters including yield and is presented in Table 8.22. The standard error of mean was found to be within a range of 0.024 to 0.147 for all the biometric parameters.

TABLE 8.21 Analysis of Variance showing Interaction Effects Among Treatments and Irrigation Levels

Dependent variable	Sum of Squares	df	Mean Square	F	Sig.
Plant height at active tillering stage	94.567	9	10.51	284.95**	.000
Plant height at panicle initiation stage	150.274	9	16.70	913.35**	.000
Plant height at flowering stage	420.960	9	46.77	1645.00**	.000
Plant height at maturity	104.097	9	11.57	429.13**	.000
Number of tillers after 30 DAP	7.781	9	0.87	1.46NS	.245
Number of tillers after 40 DAP	3.781	9	0.42	1.03NS	.456
Number of tillers after 50 DAP	26.750	9	2.97	4.32*	.005
Root length at active tillering stage	9.880	9	1.10	20.54NS	.000
Root length at panicle initiation stage	7.979	9	0.89	16.69**	.000
Root length at flowering stage	12.931	9	1.44	27.37**	.000
Root length at maturity	10.126	9	1.13	47.65**	.000
Number pf panicle/hill	30.625	9	3.40	7.78**	.000
Panicle length in cm	12.398	9	1.38	35.97**	.000
Test weight, g	7.385	9	0.82	10.26**	.000
Yield, q/h	33.548	9	3.73	114.25**	.000
Dry straw yield, q/h	54.958	9	6.11	165.47**	.000

* Significant at 1% level. ** Significant at 5% level. NS = non-significant; df = degree of freedom.

8.4.3 WATER PRODUCTIVITY AND COST-EFFECTIVENESS OF MICRO IRRIGATION METHODS IN SRI RICE CULTIVATION

8.4.3.1 Water Productivity

Some researchers have reported that 5 m³ of water is required to produce 1 kg of rice in standing water paddy. This gives the water productivity of rice as 0.2 kg/m³ which is very low and non-profitable. But in case of SRI with drip and plastic mulch, the productivity was estimated to be 0.7 kg/m³ followed by SRI with drip (0.64 kg/m³), SRI with micro sprinkler (0.54 kg/m³) and SRI with controlled irrigation (0.37 kg/m³). It was also revealed that the productivity of hybrid rice crop was 86.15 q/ha at 10% MAD level under SRI with drip and plastic mulch and it was followed by SRI with drip (83.41 q/ha), SRI with micro sprinkler (80.60 q/ha) and SRI

TABLE 8.22 Statistical Analysis of Grand Mean of Yield and Yield Attributes

Dependent variable	Mean	Std. error	95% Confidence interval	
			Lower bound	Upper bound
Plant height at active tillering stage	37.475	0.034	37.403	37.547
Plant height at panicle initiation stage	48.706	0.024	48.656	48.757
Plant height at flowering stage	72.847	0.030	72.784	72.910
Plant height at maturity	101.003	0.029	100.941	101.064
Number of tillers after 30 DAP	13.781	0.136	13.492	14.070
Number of tillers after 40 DAP	21.156	0.113	20.917	21.395
Number of tillers after 50 DAP	33.750	0.147	33.439	34.061
Root length at active tillering stage	16.053	0.041	15.966	16.140
Root length at panicle initiation stage	18.856	0.041	18.770	18.943
Root length at flowering stage	21.419	0.041	21.333	21.505
Root length at maturity	26.427	0.027	26.369	26.484
Number of panicle/hill	27.812	0.117	27.565	28.060
Panicle length in cm	27.057	0.035	26.984	27.130
Test weight, g	35.187	0.050	35.082	35.293
Yield, q/h	72.836	0.032	72.768	72.903
Dry straw yield, q/h	33.702	0.034	33.630	33.774

with controlled irrigation (52.60 q/ha). From the above information it was realized that the water productivity in SRI with drip and plastic mulch was 11% more than that of SRI with drip at 10% MAD level (Figure 8.10).

8.4.3.2   Cost-Effectiveness of Micro Irrigation

The data from this experiment revealed that use of different types of micro irrigation system in SRI method of rice cultivation increased the grain yield significantly as compared to SRI with controlled irrigation. Also the straw yield was higher in case of SRI with different types of micro irrigation method. The cost of cultivation for all the four treatments is presented in Table 8.23.

A comparison of net return per hectare showed that the increase in grain yield had a marked impact on farmers' net return as evidenced from SRI

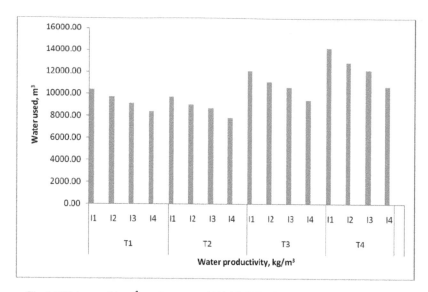

Fig. (a) Water used in m³ per hectare at 10% MAD level under different treatments.

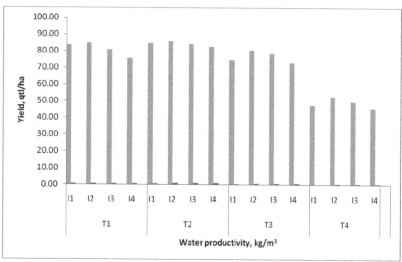

Fig. (b) Water productivity in kg/m³ 10% MAD level under different treatments

FIGURE 8.10   Water used in m3 per ha (top) and water productivity (bottom) at 10% MAD level under different treatment.

TABLE 8.23   Benefit–Cost Analysis of SRI Method of Rice Cultivation under Different Irrigation Practices

	T1	T2	T3	T4
Cultural operations	**Rs. (One US\$ = 60.00 Indian Rupees)**			
1. Land preparation, Rs.	180.00	180.00	180.00	125.00
2. Installation of micro irrigation system for an experimental area of 320 sq.m (5 year life span of micro irrigation system, 2 crops per year), Rs.	490.00	490.00	410.00	-
3. Mulching sheet cost with spreading, Rs.	-	800.00	-	-
4. Weedicide application, Rs.	45.00	45.00	45.00	45.00
5. Seedling development, Rs.	100.00	100.00	100.00	100.00
6. Transplanting, Rs.	400.00	400.00	400.00	400.00
7. Growth promoter (Biozyme), Rs.	35.00	35.00	35.00	35.00
8. Manual weeding, Rs.	200.00	-	250.00	150.00
9. NPK Fertilizer with FYM, Rs.	320.00	320.00	320.00	320.00
10. Cost of harvesting (Manual), Rs.	40.00	40.00	40.00	40.00
11. Total cost of cultivation per experimental plot, Rs.	1810.00	2410.00	1780.00	1215.00
12. Total cost of cultivation, Rs per ha.	56562.50	75312.50	55625.00	37968.75
13. Yield in quintal from the experimental plot	2.44	2.53	2.30	1.47
14. Yield in quintal per ha.	76.25	84.40	76.80	48.90
15. Selling price of paddy, Rs. per quintal	1310	1310	1310	1310
16. Realization, Rs.per ha.	106437.50	110564.00	100608.00	64059.00
17. Net profit, Rs. per ha.	49875.00	35251.50	44983.00	26090.25
18. B:C Ratio	1.88	1.47	1.81	1.69

method of rice cultivation with micro irrigation methods. The net return per hectare in SRI with drip irrigation was comparatively higher than the SRI method of rice cultivation with controlled irrigation. The benefit–cost ratio in case of SRI with drip irrigation was found to be the highest over other treatments. If cultivation of rice is made in larger scale, the cost of cultivation will reduce further, for which B:C ratio may be increased substantially.

From the above data it is clearly revealed that the net return per hectare with SRI and drip irrigation was comparatively higher by an amount of Rs. 23,784.75 which was 91.1% higher than the SRI method of rice cultivation with controlled irrigation. Although the realization per ha in case of SRI with drip and plastic mulch was the highest, i.e., Rs. 110,564.00, but the net return per ha was less in comparison to SRI with drip irrigation because of the additional cost associated with the plastic mulching sheet.

8.5   CONCLUSIONS

The effects of different micro irrigation practices on yield and yield attributes of SRI method of hybrid rice cultivation were studied and the results were discussed vividly in the results and discussion part of the research work. The following conclusions were drawn from the study:

a. Irrigation water requirement in case if SRI with drip and plastic mulch was found to be the lowest at 30% MAD level whereas, it was found to be the highest in case of SRI with controlled irrigation.

b. Total irrigation requirement of rice in SRI with controlled irrigation at 0% MAD level was found to be 66.7%, 56.53% and 31.52% more as compared to SRI with drip and plastic mulch, SRI with drip irrigation and SRI with micro sprinkler, respectively.

c. There was a significant saving of 42% water in SRI with drip and plastic mulch and 34 % water in SRI with drip irrigation at 10% MAD level as compared to SRI with controlled irrigation.

d. SRI with controlled irrigation at 10% MAD level recorded the highest average plant height at all growth stages and it was found to be superior over all other treatments.

e. From analysis of variance, it was found that the variation between treatments and variation between irrigation levels with regard to plant height at all stages was significant.

f. SRI with micro sprinkler at 10% MAD level recorded the maximum number of effective tillers per hill, which was found to be superior over other treatments followed by SRI with drip, SRI with drip and plastic mulch and SRI with controlled irrigation.

g. Average number of effective tillers was found to be more in SRI with micro sprinkler than the other irrigation practices.

h. Comparatively higher root length was found in SRI method of cultivation with controlled irrigation as compared to all other methods of irrigation at all MAD levels.

i. The maximum average panicle length was recorded in SRI method of cultivation with drip and the lowest was recorded in SRI with controlled irrigation.

j. SRI with drip irrigation at 10% MAD level recorded the maximum number of panicles and found to be superior over all other treatments followed by SRI with drip and plastic mulch, SRI with micro sprinkler and SRI with controlled irrigation.

k. SRI with drip irrigation at 10% MAD level recorded the maximum test weight of 1000 grains, which was comparatively superior over all other treatments followed by SRI with drip and plastic mulch, SRI with micro sprinkler and SRI with controlled irrigation, respectively.

l. SRI with drip irrigation and plastic mulch at 10% MAD level recorded the maximum yield.

m. Highest straw yield was also obtained in SRI with drip and plastic mulch as compared to other irrigation practices.

n. The interaction effect of treatment and irrigation levels was found to be significant in case of plant height, panicle length, number of panicles, test weight, yield and dry straw yield at all growth stages was found to be significant at 1% level of significance. The root length at all stages except tillering stage was found to be significant at same level of significance.

o. The water productivity of hybrid rice crop in SRI with drip and plastic mulch was found to be the highest at 10% MAD level followed by SRI with drip, SRI with micro sprinkler and SRI with controlled irrigation.

p. The benefit–cost ratio was found to be more in case of SRI with drip as compared to other irrigation practices.

8.6 SUMMARY

Currently, mainstream technological options to improve rice production focus mainly on selection of improved varieties, proper crop nutrition and weed management, pest and diseases control and irrigation management. Interestingly, SRI is an alternative technique that depends on such agronomic practices with less input requirement and maximizes yield. Studies on water-saving technologies in rice production systems have mainly focused on the innovations in cultivation systems that incorporate furrow irrigation, sprinkler irrigation, drip irrigation and drip irrigation under plastic mulching which gave some inspiring records of water-saving for rice cultivation. In recent times, it is very important to increase the productivity, water use efficiency (WUE) and production of rice crop under the drip irrigation system under water scarce situation.

The present study was undertaken (i) to estimate water requirement of rice crop grown by SRI method under various irrigation practices, (ii) to compare yield and yield attributes of rice at different irrigation levels and (iii) to study the economics of different irrigation systems on rice crop. The water productivity of hybrid rice crop in SRI with drip and plastic mulch was found to be the highest at 10% MAD level followed by SRI with drip, SRI with micro sprinkler and SRI with controlled irrigation. The benefit–cost ratio was found to be more in case of SRI with drip as compared to other irrigation practices.

KEYWORDS

- agronomic practice
- available soil moisture content
- benefit–cost ratio
- cost of cultivation
- drip irrigation
- dripper field capacity
- furrow irrigation

- **grain weight**
- **irrigation management**
- **irrigation method**
- **irrigation practice**
- **management allowable deficit**
- **net profit**
- **panicle**
- **plant height**
- **plastic mulch**
- **productivity**
- **reproductive stage**
- **rice**
- **root length**
- **saturation**
- **sprinkler irrigation**
- **tiller**
- **water management**
- **water productivity**
- **water saving irrigation**
- **water use efficiency**
- **yield**
- **yield attributing factors**

REFERENCES

1. Anonymous. (2010). *Handbook of Statistics on the Indian Economy.* Government of India, New Delhi, pp. 135.
2. Cheng, J. P., Cao G., & Cai, M. L. (2006). Effects of different irrigation modes on the yield and water productivity of rice. *Chin. Soc. Trans. Agric. Eng.*, *22*(11), 28–32.
3. Geethalakshmi, V., Lakshmanan, A., & Ramesh, T. (2008). Assessment of water requirement for different systems of rice cultivation. *International Journal of Research in Biosciences, 2*(2), 1–12.
4. Govindan, R., & Grace, T. M. (2012). Influence of drip fertigation on growth and field of rice varieties. *Madras Agricultural Journal*, *99*(4/6), 244–247.

5. Haibing, H., Fuyu, M., Ru Yang, L. C., Biao, J., Jing, C., Hua, F., Xin, W., & Li, L. (2013). Rice performance and water use efficiency under plastic mulching with drip irrigation. *PLoS One, 8*(12), 102–118.

6. He, H. B., Yang, R., Chen, L., Fan, H., Wang, X., Wang, S. Y., Cheng, H. W., & Ma, F. Y. (2014). Rice root system spatial distribution characteristics at flowering stage and grain yield under plastic mulching drip irrigation (PMDI). *The Journal of Animal & Sciences, 24*(1), 290–301.

7. Jiang, T, Feng, H., Hui-xin, L., Yi-ping, W., Fa-Quan, H., & Hua-Xiang, H. (2006). Effects of non-flooded cultivation with straw mulching on rice agronomic traits and water use efficiency. *Rice Science, 13*(1), 59–66.

8. Islam, M. S., Bhuiya, M. S. U., Rahman, S., & Hussain, M. (2009). Evaluation of Spad and LCC based nitrogen management in rice (*Oryzasativa* L.). *Bangladesh J. Agril. Res., 34*(4), 661–672.

9. Kumar, V., Gurusamy, A., Mahendean, P. P., & Mahendran, S. (2009). Optimization of water and nutrient requirement for yield maximization in hybrid rice under drip fertigation system. *Rice Science, 23*(1), 55–64.

10. Lal, B., Gautam, P., & Joshi, E. (2013). Different rice establishment methods for producing more rice per drop of water: A review. *International Journal of Research in Biosciences, 2*(2), 1–12.

11. Liu, X., Li Qiao, L., Lu, Z., Feng, D., Li, P., Fan, X., & Xu, K. (2013). Selection of thickness of high-density polyethylene film for mulching in paddy rice. *American Journal of Plant Sciences, 4*, 1359–1365.

12. Ali, M., Ahmad, R., Ahmad, Z., & Muhammad, A. S. (2013). Effect of different mulches techniques on weed infestation in aerobic rice (*Oryza sativa* L.). *American Eurasian J. Agric. & Environ. Sci., 13*(2), 153–157.

13. Panigrahi, B., Roy, D. P., & Panda, S. N. (2010). Water use and yield response of tomato as influenced by drip and furrow irrigation. *International Agricultural Engineering Journal, 19*, 1–19.

14. Patra, P. S., & Haque, S. (2011). Effect of seedling age on tillering pattern and yield of rice (*Oryza sativa* L.) under system of rice intensification. *ARPN Journal of Agricultural and Biological Science, 6*(11), 175–184.

15. Prasanna, P. A., Lakshmi, K. S., & Singh, A. (2009). Rice production in India-implications of land inequity and market imperfections. *Agricultural Economics Research Review, 22*, 431–442.

16. Rao, K. V. R. (2013). Evaluation of drip irrigation system in paddy crop: a viable alternate to conventional water management practice in paddy cultivation. *International Exhibition and Conference on Water Technologies, Environmental Technologies and Renewable Energy*, Bombay Exhibition Center, Mumbai, India.

17. Singh, R. (2005). Influence of mulching on growth and yield of tomato (*Solanum Lycopersicum* L.) in north Indian plains, *Veg. Sci., 32*(1), 55–58.

18. Qamar, R., Kalim, M., Rehman, A., Iqbal, Z., Ghaffar, A., & Mustafa, G. (2014). Growth and economic assessment of mulches in aerobic rice (*Oryza Sativa* L.). *J. Agric. Res., 52*(3), 395–406.

19. Yadav, G. S., Shivay, Y. S., Kumar, D., & Babu, S. (2013). Enhancing iron density and uptake in grain and straw of aerobic rice through mulching and rhizo-foliar fertilization of iron. *African Journal of Agricultural Research, 8*(10), 5447–5454.

CHAPTER 9

WATER USE EFFICIENCY FOR SUGARCANE UNDER DRIP IRRIGATION

R. GUPTA

CONTENTS

9.1   INTRODUCTION

Water is a scarce natural resource. Its per capita availability is decreasing due to climate change and increased population, and has already reached to water stress line. Availability of water for agriculture is further decreasing as more and more quantity of water is being diverted for developmental activities and industrialization. With the conventional irrigation practices (small and long borders, furrow irrigation and trench irrigation), a significant amount of water is lost in evaporation from the land surface especially during the initial stages of crop growth when the actual canopy cover is much less. Secondly, the return flows from irrigation do not percolate deep to join the groundwater table. The vadose zone holds water moving vertically downwards as hygroscopic water and capillary water. All this water would eventually get evaporated from land after the crop harvest.

Drip irrigation system considerably reduces the wetted land surface and applies water in such a quantity that it remains within root zone of the plants thereby making evaporation and percolation losses nearly zero. The Indian sugar industry is striving to increase efficiency of its resource use and reduce off-site impacts of production. One option for increasing resource use efficiency and mitigating the effect of climate change in irrigated sugar production systems is the adoption of drip irrigation [5, 6, 9]. It is commonly accepted that the efficiency of fertilizer use can be improved when it is applied by fertigation to sugarcane [7].

An experiment was conducted by Pawar et al. [8] who reported that drip irrigation used less quantity of water (103.7 mm) and saved 57% water over surface irrigation method. They also observed that 80% drip fertigation through water soluble fertilizers applied in 26 splits as per growth stages is more suitable for productive sugarcane cultivation in western Maharashtra. As reported by Bhunia et al. [2], crop geometry under drip irrigation system significantly influenced cane yield and highest yield of 130.64 t/ha was recorded with 75 cm row spacing followed by 126.13 t/ha at 90 cm row spacing. They attributed the higher yield in 75 and 90 cm row spacing to higher tillers/m^2, cane length and inter node length. Paired row plantings fetched lower yields where single drip line was kept between two paired rows of sugarcane. While there are recommendations for N application rates in conventional management systems [3], there is little information on what the optimum N rate should be fertigated sugarcane [1, 9].

The present research was initiated to compare the performance of drip irrigation with surface irrigation for sugarcane crop.

9.2   METHODS AND MATERIALS

9.2.1   EXPERIMENTAL SITE AND CLIMATE

Field experiments were conducted during 2006 – 2010 at research farm of the Indian Institute of Sugarcane Research, Lucknow (26°56′N–80°52′E and 111 m above mean sea level). Climate of the experimental site is semi-arid, subtropical with hot dry summers and cold winters. The average annual rainfall is 976 mm, and nearly 80% of the rainfall is received from southwest monsoons during July–September. Average monthly minimum temperature is 7.5°C in January (the coldest month) and 27°C in May (the hottest month). The respective average maximum temperatures are 22.1°C in January and 39.8°C in May. Experiment was conducted on a silty clay loam soil of Indo-Gangetic alluvial origin; very deep (>2 m), well drained, flat and classified as non-calcareous mixed *hyperthermic Udic Ustochrept*. Initial soil texture analysis showed that the soil texture was silty clay loam with 38.4% sand, 34.2% silt and 27.4% clay. Bulk density of the soil was 1.35 g/cc. Soil moisture at field capacity and permanent wilting point were 27% and 9.6% respectively. The soil had a pH of 7.60 with 0.50% organic carbon (OC) and 260, 43 and 353 kg of available N, P_2O_5 and K_2O ha^{-1}, respectively.

9.2.2 CULTURAL PRACTICES

Sugarcane was planted in the first week of February in the year 2007 and again in 2008 at 75 cm row spacing with a seed rate of 45,000 three budded cane sets per ha. Minimum plot size was kept to 180 m × 7.5 m (1350 m^2). Following the main sugarcane plant crop, a ratoon crop was initiated in the first week of February and the same treatments were imposed. Of the total recommended dose (NPK −150, 60, 60 kg/ha for plant cane and 200, 60, 60 kg/ha for ratoon), 1/3rd of the N and the full amount of P and K were applied at the time of planting and the remaining 2/3rd of the N was applied through weekly fertigation in drip irrigated plots and top dressed in two installments in surface irrigated plots during the tillering phase from April to June. After each irrigation, intercultural operations were done in the ratoon crop also. Furadan 3G@1.0 kg active ingradient/ha was applied before onset of monsoon to avoid incidence of top borer. The harvesting of sugarcane plant and ratoon crops was done manually in the last week of January close to the ground level using specially designed steel choppers. Following the plant crop, transplanting of clumps was done in gaps created due to mortality of clumps to have a uniform plant population.

9.2.3　TREATMENTS

Irrigation was scheduled using a meteorological approach based upon pan evaporation. In drip method of irrigation, irrigation water equivalent to 0.6, 0.8, 1.0 and 1.2 times pan evaporation was applied at 2 days interval, whereas in surface irrigation, 80 mm irrigation water was applied when cumulative pan evaporation is 80 mm (IW:CPE = 1.0 where IW is irrigation water applied and CPE is cumulative pan evaporation). All the irrigation treatments were replicated thrice. Irrigation water use efficiency (IWUE) was calculated as follows:

$$\text{Cane yield (kg/ha)} = \text{Total irrigation water use (mm)} \qquad (1)$$

9.2.3　STATISTICAL ANALYSIS OF DATA

The data were statistically analyzed and treatments were compared sing randomized block design (RBD) for each crop season. The critical difference (CD) is defined as the least significant difference beyond which all treatment differences were statistically significant, and it was computed by the following formula.

$$\text{C.D.} = (\sqrt{2VE_r})X_{t5\%} \qquad (2)$$

where, VE is the error variance, r is number of replications, $t5\%$ the table value of t at 5% level of significance at error degree of freedom [4].

9.3　RESULTS AND DISCUSSION

9.3.1　CROP GROWTH PARAMETERS

The crop growth parameters viz. number of tillers before onset of monsoon, number of millable canes, tiller mortality, cane stalk diameter and cane stalk length are presented in Table 9.1 for plant crop and in Table 9.2 for ratoon crop. It has been observed that all these parameters are influenced significantly by irrigation treatments. Drip irrigation resulted in higher number of tillers (Figure 9.1) and millable canes than that from surface irrigated crop. It was observed that tiller mortality was higher in the treatments which gave higher number of tillers before onset of monsoon. Competition for nutrients, air, water and light due to overcrowding of tillers at one place might have resulted in tiller mortality. Tiller mortality both for plant (53.6%) and ratoon (63.1%) crops was the highest in drip irrigated sugarcane when applied irrigation water was equal to 0.8 times of pan evaporation. However, tiller mortality and cane stalk diameter for ratoon crop were not significantly influenced by irrigation treatments. Cane stalk diameter and length were higher in drip-irrigated crop as compared to surface irrigated crop. Irrigation water equal to 0.8 times pan evaporation resulted in the highest cane stalk diameter (2.39 mm for plant and 2.35 mm for ratoon crops respectively) and cane stalk length (247.8 cm for plant and 242.4 cm for ratoon crops, respectively).

FIGURE 9.1   Variation in tillering pattern of drip and surface irrigated crop.

TABLE 9.1 Effects of Different Irrigation Treatments on Number of Tillers at Onset of Monsoon, Number of Millable Canes, Tiller Mortality, Cane Stalk Diameter and Cane Stalk Length for Plant Crop

Fertigation treatments	Number of tillers before onset of monsoon	Number of millable canes	Tiller mortality (%)	Cane stalk diameter (cm)	Cane stalk length (cm)
Drip irrigation at 2 days interval with irrigation water equal to 0.6 E_{pan}	183,800	86,333	53.03	2.25	229.4
Drip irrigation at 2 days interval with irrigation water equal to 0.8 E_{pan}	197,400	91,600	53.60	2.39	247.8
Drip irrigation at 2 days interval with irrigation water equal to 1.0 E_{pan}	190,233	92,800	51.22	2.33	242.2
Drip irrigation at 2 days interval with irrigation water equal to 1.2 E_{pan}	160,700	89,733	44.16	2.32	229.3
Conventional flood irrigation with 8 cm irrigation water at 1.00 IW/CPE ratio	170,967	82,367	51.82	2.21	217.9
SE mean ±	1939	1330	0.65	0.03	3.1
CD 0.05	4471	3068	1.50	0.06	7.1

9.3.2 SUGARCANE YIELD AND IRRIGATION WATER USE EFFICIENCY

The effect of irrigation treatments on sugarcane yield and IWUE was statistically significant for plant as well as ratoon crop (Tables 9.3 and 9.4). Drip irrigation resulted in higher sugarcane yield as well as higher IWUE. For plant crop, the highest yield of 93.48 t/ha was recorded, when fertigation was done with the amount of water equal to 0.8 times pan evaporation. With this fertigation treatment, IWUE was 1451.3 kg/ha-cm. The lowest cane yield (69.74 t/ha) was obtained in surface irrigation treatment with 866.1 kg/ha-cm of IWUE. The highest IWUE of 1647.1 kg/ha-cm was recorded when fertigation was done with the amount of water equal to 0.6

TABLE 9.2   Effects of Irrigation Treatments on Number of Tillers at Onset of Monsoon, Number of Millable Canes, Tiller Mortality, Cane Stalk Diameter and Cane Stalk Length for Ratoon Crop

Fertigation treatments	Number of shoots before onset of monsoon	Number of mill-able canes	Tiller mortality (%)	Cane stalk diameter (cm)	Cane stalk length (cm)
Drip irrigation at 2 day interval with irrigation water equal to 0.6 E_{pan}	214,444	89,400	58.31	2.25	228.7
Drip irrigation at 2 day interval with irrigation water equal to 0.8 E_{pan}	254,583	93,967	63.09	2.35	242.4
Drip irrigation at 2 day interval with irrigation water equal to 1.0 E_{pan}	234,722	97,067	58.65	2.34	241.8
Drip irrigation at 2 day interval with irrigation water equal to 1.2 E_{pan}	231,111	91,500	60.41	2.28	231.6
Conventional flood irrigation with 8 cm irrigation water at 1.0 IW/CPE ratio	207,500	82,700	60.14	2.17	219.6
SE mean ±	13218	13218	NS	NS	6.4
CD 0.05	30481	30481			14.8

times pan evaporation. At this IWUE, cane yield was 79.6 t/ha. The yield difference between fertigation treatments receiving irrigation water equal to 0.6 and 0.8 times the pan evaporation was found to be non-significant but the difference in IWUE was statistically significant.

For ratoon crop, highest cane yield of 91.36 t/ha was recorded when fertigation was done with the amount of water equal to 1.0 times pan evaporation. For this fertigation treatment, IWUE was 1134.6 kg/ha-cm. The yield difference between fertigation treatments receiving irrigation water equal to 0.8 and 1.0 times the pan evaporation was found to be non-significant but the difference in IWUE was statistically significant. Lowest cane yield (62.35 t/ha) was obtained in surface irrigation treatment with 774.3 kg/ha-cm IWUE. Highest IWUE of 1498.2 kg/ha-cm

TABLE 9.3   Effects of Irrigation Treatments on Irrigation Water Depth, Sugarcane Yield, Irrigation Water Use Efficiency for Plant Crop

Fertigation treatments	Irrigation water applied (mm)	Yield (t/ha)	IWUE (kg/ha-cm)
Drip irrigation at 2 days interval with irrigation water equal to 0.6 E_{pan}	483.1	79.57	1647.07
Drip irrigation at 2 days interval with irrigation water equal to 0.8 E_{pan}	644.1	93.48	1451.33
Drip irrigation at 2 days interval with irrigation water equal to 1.0 E_{pan}	805.2	92.31	1146.42
Drip irrigation at 2 days interval with irrigation water equal to 1.2 E_{pan}	966.2	83.22	861.31
Conventional flood irrigation with 8 cm irrigation water at 1.00 IW/CPE ratio	805.2	69.74	866.12
SE mean ±		5.08	71.8
CD (0.05)		11.70	165.6

was recorded when fertigation was done with the amount of water equal to 0.6 times pan evaporation. For this fertigation treatment, cane yield was 72.4 t/ha.

9.4   CONCLUSIONS

Results indicate that drip irrigation improved number of millable canes, and diameter and length of cane stalk which in turn resulted in higher sugarcane yield. Drip irrigation also improved IWUE significantly due to saving in the irrigation water and higher sugarcane yield. Depending on the amount of irrigation water applied, sugarcane yield of drip irrigated crop was improved by 14–34% for plant crop and 16–46% for ratoon crop. IWUE also improved by 32–90% for plant crop and by 18–93% for ratoon crop. With the same quantity of fertilizers and reduced amount of irrigation water, higher sugarcane yield is achievable. The results thus suggest that drip irrigation not only enhance irrigation water and nitrogen use efficiency but also enhance sugarcane yield.

TABLE 9.4 Effects of Irrigation Treatments on Irrigation Water Depth, Sugarcane Yield, Irrigation Water Use Efficiency for Ratoon Crop

Fertigation treatments	Irrigation water applied (mm)	Yield (t/ha)	IWUE (kg/ha-cm)
Drip irrigation at 2 days interval with irrigation water equal to 0.6 E$_{pan}$	483.1	72.38	1498.24
Drip irrigation at 2 days interval with irrigation water equal to 0.8 E$_{pan}$	644.1	90.47	1404.60
Drip irrigation at 2 days interval with irrigation water equal to 1.0 E$_{pan}$	805.2	91.36	1134.62
Drip irrigation at 2 days interval with irrigation water equal to 1.2 E$_{pan}$	966.2	88.57	916.68
Conventional flood irrigation with 8 cm irrigation water at 1.00 IW/CPE ratio	805.2	62.35	774.34
SE mean ±		2.24	23.1
CD (0.05)		5.16	53.4

9.5 SUMMARY

Water and nitrogen are two most essential inputs for achieving maximum yields of sugarcane. Balanced use and application of these critical inputs not only help in sustaining sugarcane productivity but also enhance the profitability of the farmer. Application of excess water not only results in wastage but can also take away precious nutrients. On the other hand application of excess nitrogen will affect profitability of farmer. On the contrary, inadequate application of water and nitrogen will affect the crop yield and profitability adversely. To utilize these two major resources optimally, experiments were conducted at Indian Institute of Sugarcane Research, Lucknow, to compare sugarcane crop performance under drip and surface irrigation.

Results indicate that for plant crop, the highest cane yield of 93.48 t/ha was recorded when fertigation was scheduled at 0.8 Epan. At this fertigation scheduling, IWUE was 1451.3 kg/ha-cm. For plant crop, the lowest cane yield (69.74 t/ha) was obtained in surface irrigation treatment with 866.1 kg/ha-cm IWUE. Highest IWUE of 1647.1 kg/ha-cm was recorded when fertigation was scheduled at 0.6 Epan. At this IWUE, cane yield was 79.6 t/ha.

For ratoon crop, highest cane yield of 91.36 t/ha was recorded when ferti-gation was scheduled at 1.00 Epan. At this fertigation scheduling, IWUE was 1134.6 kg/ha-cm whereas the lowest cane yield (62.35 t/ha) for ratoon was obtained in surface irrigation treatment with 774.3 kg/ha-cm IWUE. For ratoon crop, the highest IWUE of 1498.2 kg/ha-cm was recorded when fertigation was scheduled at 0.6 Epan. At this IWUE, cane yield was 72.4 t/ha. The results suggest that drip irrigation not only enhance irrigation water and nitrogen use efficiency but also enhance sugarcane yield.

KEYWORDS

- cane stalk diameter
- cane stalk length
- critical difference
- error variance
- fertigation
- irrigation water
- irrigation water use efficiency
- millable canes
- pan evaporation
- randomized block design
- ratoon
- sugarcane
- surface irrigation
- tiller mortality
- treatment
- yield

REFERENCES

1. Bar-Yosef, B. (1999). Advances in fertigation. *Adv Agron, 65*,102–110.
2. Bhunia, S. R., Chauhan, R. P. S., & Yadav, B. S. (2013). Effect of planting geom-etry and drip irrigation levels on sugarcane (*Saccharum officinarum*) yield, water

use efficiency and cane quality under transgenic plants of Rajasthan. *J. Progressive Agri., 4*(1), 102–104.

3. Calcino, D. V. (1994). *Australian Sugarcane Nutrition Manual. BSES/SRDC, Brisbane,* Australia, 28 pp.

4. Cochran, W. G., & Cox, G. M. (1957). *Experimental Designs.* Second edition. John Willey, Sons, New York, USA, 611 pp.

5. Keating, B. A., Verburg, K., Huth N. I., & Robertson, M. J. (1997). Nitrogen management in intensive agriculture: sugarcane in Australia. In: Keating, B.A., Wilson, J. R. (eds.). *Intensive Sugarcane Production: Meeting the Challenge Beyond 2000, CAB International,* Wallingford, UK, p. 221.

6. Meyer, W. S. (1997). The irrigation experience in Australia: lessons for the sugar industry. In: Keating, B. A., & Wilson, J. R. (eds.). *Intensive Sugarcane Production: Meeting the Challenge Beyond 2000, CAB International,* Wallingford, UK, p. 437.

7. Ng Kee Kwong, K. F., & Deville, J. (1994). Application of ^{15}N-labeled urea to sugarcane through a drip irrigation system in Mauritius. *Fert. Res., 39,* 223–228.

8. Pawar, D. D., Dingre, S. K., & Surve, U. S. (2013). Growth, yield and water use in sugarcane (*Saccharum officinarum*) under drip fertigation. *Indian J. Agronomy, 58*(3), 396–401.

9. Thorburn, P. J., Sweeney, C. A., & Bristow, K. L. (1998) Production and environmental benefits of trickle irrigation for sugarcane: a review. *Proc. Aust. Soc. Sugar Cane Technol., 20,* 118–125.

CHAPTER 10

WATER USE EFFICIENCY OF SORGHUM UNDER DRIP IRRIGATION

U. M. KHODKE

CONTENTS

10.1   INTRODUCTION

Sorghum *(Sorghum bicolarmonech)* is the third main crop in terms of cultivated area next to paddy and wheat in semi-arid and arid climate zones of central and southern India. It is the most vital food and fodder crop grown in *kharif* (rainy), *rabi* (post rainy or winter) and summer seasons. *Rabi* sorghum covers an area of 3.5 M-ha alone in Maharashtra [3] with an equal area in adjoining Karnataka and Andhra Pradesh states. Despite

---

In this chapter: 1 q (quintal) = 100 kg.

slight decrease in the cultivated area of *rabi* sorghum over the years, the production level during 2003 in India was almost similar to that in the early 1970's, which could be largely attributed to adoption of improved varieties and hybrids [5]. However, over the past two decades, the yields of sorghum are relatively static, with productivity not having improved.

Rabi sorghum has its own value due to better grain and fodder quality. *Rabi* sorghum mostly is grown under rain-fed conditions with some protective irrigation if available. However the area under *rabi* sorghum in the region is decreasing due to nonoccurrence of late monsoon or early winter rains resulting in to unavailability of soil moisture. Yield and quality of sorghum crop often suffers due to presence of insufficient soil moisture during its growth period. The water stress during the critical growth stages of sorghum necessitates the efficient use of irrigation water so as to minimize significant loss of crop yields. Non-availability of water during rabi season in semi-arid regions of Maharashtra (India) is one of the major constraints that limits the productivity of sorghum. Traditionally *rabi* sorghum is grown as rain-fed crop in *kharif* and conventional irrigation method fails to supply the adequate quantity of water at proper time.

The seasonal ET_C of sorghum under rain-fed conditions ranged from 168.13 to 245.19 mm. Gundekar and Khodke [4] developed crop coefficients for *rabi* sorghum using ETo (reference crop evapotranspiration) derived from FAO 56 procedure and actual water use of the crop experimentally measured using lysimeters under rain-fed semi-arid conditions of Parbhani. This average measured seasonal ETc amounting to 218.5 mm was considerably lower than those reported for studies conducted in similar climatic regions, ranging from 225 to 388 mm (August planting) at Bellary [7] in semi-arid tropics and 490 to 500 mm (June planting) at Hisar [9] in arid tropics of India. In another study in semi-arid tropical zone of Nigeria, *ETc* of sorghum on the basis of 10-day water balance was calculated [1]. It was assumed that although the lysimeter was not irrigated, yet sufficient rainfall was received to make crop water use data valid.

Under the situations of non-availability of soil moisture during dry spell in semi-arid regions of the country, drip irrigation can prove be beneficial even for close growing crops due to its multifarious advantages. Studies reported in the literature reveal that with the use of drip, there are water savings as well as increase in yield [2, 6]. Now-a-days drip irrigation system is becoming popular among the farmers in India under close growing and high value crops [8].

This chapter assesses the maximum yield potential of *rabi* sorghum under drip irrigation and to test the promising varieties of *rabi* sorghum which responds better to irrigation.

10.2 MATERIALS AND METHODS

The field experiments were conducted at Water Management Farm of Vasantrao Naik Marathwada Agricultural University Parbhani–India in factorial randomized plot design with irrigation levels and sorghum varieties as factors. Experiment comprised of five irrigation scheduling treatments with three replications:

I_1 drip at 100% ETc (crop evapotranspiration),

I_2 drip with 75% ETc,

I_3 drip at 100% ETc during critical growth stages only (grand growth, flag leaf, flowering and grain filling) and compared with,

I_4 border check basin irrigation at 0.8 IW/ CPE, and

I_5 rain-fed (control).

The surface irrigation was scheduling when cumulative pan evaporation (CPE) reached 75 mm with depth of irrigation (IW) as 60 mm. Varieties tested were Parbhani Jyoti (SPV-1595) and Akola Kranti (SPV-1549), which respond well to irrigation tested.

The soil at the experimental plot is low in organic carbon and available nitrogen, medium in phosphorus, fairly rich in potassium and slightly alkaline in reaction. The crop coefficient curves for *rabi* sorghum were developed by following the guidelines of FAO-56 and the ETc was estimated using reference crop evapotranspiration and crop coefficient. The daily pan evaporation data collected from the Indian Meteorological Department (IMD) observatory located near the experimental field was used.

The plants were placed at 15 cm in a crop row and distance between two rows was 45 cm. The inline 16 mm diameter drip laterals having drippers of 2.54 lph discharge and 30 cm spacing were laid for two rows of sorghum placed at 45 cm apart. The distance of 75 cm distance was kept between the two pairs of rows. Hence the spacing of drip lateral was 120 cm. The surface irrigation at four critical growth stages was applied through basin placed apart at 2 m distance.

The seed treatment was given before sowing (Rizobium and Phosphorus Solubilizer Bacteria). The recommended dose of fertilizer (RDF) for *rabi* sorghum was 80:40:40 kg/ha of N, P_2O_5 and K_2O. In drip irrigated plots, N was applied in 3 splits (30, 30 and 20 kg/ha at 15, 30 and 60 days after sowing, DAS); phosphorous was applied in 2 splits (20 and 20 kg/ha at sowing and 30 DAS), whereas K was applied in three splits (20, 10 and 10 kg/ha at sowing and at 30, 60 and 75 DAS). In surface irrigated plots, N in 2 splits (40 kg/ha each at sowing and 30 DAS) and P & K at the time of sowing were applied. Water-soluble fertilizers of different grades were used for drip fertigation (19:19:19; 0:50:34 and Urea). There was no serious problem of incidence of pests and diseases. However, one spraying of *Chloropyriphos* was taken up as preventive and protective measures against shoot fly before 45 days after sowing. The biometric and yield observations were taken consecutively. Figure 10.1 shows the experimental plot of sorghum with drip irrigation system at different growth stages of the crop.

10.3   RESULTS AND DISCUSSIONS

10.3.1   GRAIN AND FODDER YIELD

Results (Table 10.1) indicated that I_1 was significantly superior over all other scheduling treatments. Drip irrigation with 100% ETc depth recorded significantly higher grain (57.67 q/ha: In this chapter 1.00 q = 100 kg) and dry fodder (88.57 q/ha) yield. Irrigation scheduling I_1 was comparable and at par with I_2 (drip at 75% ETc) as regards to grain and fodder yield of *rabi* sorghum. Lowest grain yield (15.92 q/ha) and fodder yield (35.51 q/ha) were recorded in I_5 (rain-fed) treatment. Results indicate that the grain and fodder yields under all drip irrigation scheduling's were significantly higher than that of surface irrigation (I_4) and rain-fed (I_5).

Among the varieties, results showed that the effect of variety on dry fodder yield was not significant but Akola Kranti AKSV 18 R (V_2) recorded

FIGURE 10.1 Sorghum crop irrigated with drip system (crop development stage: top); and sorghum crop irrigated with drip system (reproductive stage: bottom).

TABLE 10.1   Growth and Yield Contributing Parameters of Rabi Sorghum as affected by Different Treatments

Treatment	Plant height	Primaries/ panicle	Panicle length	Panicle girth	Grains wt/panicle	100 grain wt.	Grain yield	Fodder yield	Harvest index
	cm	No.	cm	cm	g	g	q/ha	q/ha	%
Irrigation scheduling									
I_1: drip at 1.0 ETc	290.0	76.7	21.58	25.07	138.0	3.88	57.67	88.57	39.5
I_2: drip at 0.75 ETc	277.6	73.0	20.79	23.87	132.1	3.68	53.78	85.83	38.6
I_3: drip at 0.75 Etc (CGS)	262.4	65.9	19.35	22.12	113.1	3.27	39.88	69.68	36.9
I_4: Surface control at CGS	244.1	60.9	18.60	20.31	94.0	3.01	33.49	61.46	35.6
I_5: Rain-fed	209.6	45.4	16.27	16.70	54.7	2.76	15.92	35.51	31.3
SEm ±	8.16	1.52	0.69	1.04	4.44	0.09	1.47	3.02	0.98
CD at 5%	24.20	4.51	2.06	3.11	13.17	0.28	4.39	8.96	2.92
Variety									
V_1	247.9	62.6	18.52	20.54	103.9	3.40	38.44	67.02	36.0
V_2	265.6	66.2	20.12	22.69	108.9	3.24	41.85	69.40	36.8
SEm ±	5.16	0.96	0.44	0.66	2.80	0.06	0.93	1.91	0.62
CD at 5%	15.31	2.85	1.30	1.96	8.33	0.17	2.77	NS	NS
Interactions									
SE m±	11.54	2.15	0.98	1.48	6.28	0.13	2.09	8.86	1.39
CD at 5%	NS	6.38	NS	4.40	18.63	NS	NS	NS	NS

Note: V_1 and V_2 are sorghum varieties namely *Parbhani Jyoti* and *Akola Kranti*.

In this chapter: One q (quintal) = 100 kg.

numerically higher fodder yields (69.40 q/ha) than Parbhani Jyoti SPV 1595 (67.02 q/ha).

10.3.2 GROWTH AND YIELD ATTRIBUTES OF RABI SORGHUM

The data on growth and yield characters of *rabi* sorghum (Plant height (cm), number of primaries per panicle, harvest index (%), panicle length (cm), panicle girth (cm) and grain weight per panicle (g)) indicate that the growth and yield characters of *rabi* sorghum are significantly influenced by irrigation scheduling whereas the variety have significant effect on some of the growth and yield characters. The mean plant height before harvest was significantly highest in irrigation scheduling I_1 (290 cm) and was at par I_2 (277.6 cm). Sorghum variety Akola Kranti (V_2) showed significantly higher plant height (265.6 cm) as compared to Parbhani Jyoti (247.9 cm). The harvest index of sorghum was significantly higher in I_1 (39.5%) and I_2 (38.6%) irrigation scheduling as compared to other irrigation scheduling's whereas the effect of variety on harvest index was not significant. The interaction effect of variety and irrigation scheduling was non-significant for all the growth parameters (Table 10.1).

10.3.3 WATER USE EFFICIENCY

The water use efficiency under different irrigation scheduling is presented in Table 10.2. Data reveals that the highest water use efficiency (WUE) was recorded under rain-fed plots where one post sowing irrigation was applied for germination. Among drip irrigation scheduling's, irrigation scheduling at 75% ETc recorded highest water use efficiency.

10.4 CONCLUSIONS

The drip irrigation scheduling at 100% ETc produced significantly higher grain and fodder yields of sorghum. Drip irrigation scheduling at 0.75 ETc depth gave highest water use efficiency followed by drip irrigation at 1.0 ETc and drip at Critical growth stages. Even with lesser water application

TABLE 10.2 Water Use Efficiency (kg/ha-mm) of Rabi Sorghum as Influenced by Different Irrigation Scheduling

Irrigation scheduling	Yield (q/ha)	Water applied (mm)	Effective Rainfall (mm)	Total water use (mm)	Water use efficiency (kg/ha-mm)
I_1	57.67	297.5	23.2	320.7	17.98
I_2	53.78	237.9	23.2	261.1	20.60
I_3	39.88	209.8	23.2	233.0	17.12
I_4	33.49	340.0	23.2	363.3	9.21
I_5	15.92	80.0	23.2	103.3	15.41

at 75% of ETc, the productivity of rabi sorghum can be enhanced. The commonly used drip system for any close growing crops such as vegetables can be used to maximize the yield of rabi sorghum. These drip systems have similar specifications require for *rabi* sorghum. The variety Akola Kranti AKSV 18R gave higher sorghum grain and fodder yield as compared to Parbhani Jyoti SPV 1595. However Parbhani Jyoti recorded significantly higher 100 grain weight as compared to Akola Kranti. No significant effect of interaction between irrigation scheduling and sorghum variety on grain and fodder yield was observed. On the basis of cost economics worked out for the present system, drip irrigation is proved to economically feasible for *rabi* sorghum with better water saving as compared to conventional surface irrigation system.

10.5 SUMMARY

Sorghum (*Sorghum bicolarmonech)* is an important staple food in addition to its additional advantages for the industrial use. *Rabi* sorghum has its own value due to better grain and fodder quality in India. The *rabi* sorghum area in the region is decreasing mainly due to nonoccurrence of late monsoon and early winter rains thereby reduction in soil moisture availability for sowing and crop growth. The unavailability of soil moisture during dry spells in semi-arid region justifies necessity of drip irrigation even for close growing crops like sorghum because of multifarious advantages of drip irrigation. In order to assess the maximum yield potential of *rabi* sorghum an experiment was conducted for three years comprising of treatments of drip and surface irrigation. Two promising varieties of *rabi* sorghum which responds better to irrigation were tested. The results based on the yield and economic evaluation indicated that drip irrigation scheduling at 100% ETc (crop evapotranspiration) recorded significantly higher grain and fodder yield and harvest index and was comparable with irrigation scheduling at 75% ETc. The grain yield is as much as 2.5 times higher than those plots irrigated with four surface irrigations at critical growth stages. The use of drip irrigation was found to enhance *rabi* sorghum productivity and water use efficiency of *rabi* sorghum.

KEYWORDS

- **critical growth stage**
- **crop coefficient**
- **drip irrigation**
- **drippers**
- **evapotranspiration**
- **grain weight**
- **harvest index**
- **irrigation scheduling**
- **lateral**
- **lysimeter**
- **pan evaporation**
- **panicle girth**
- **panicle length**
- **plant height**
- *rabi*
- **reference crop evapotranspiration**
- **sorghum**
- **surface irrigation**
- **water use efficiency**
- **yield**

REFERENCES

1. Abdulmumin, S., & Misari, S. M. (1990). Crop coefficients of some major crops of the Nigerian semi-arid tropics. *Agricultural Water management, 18,* 159–171.
2. Chawala, J. K., & Narda, M. K. (2001). Economics of fertilizer use in trickle fertigated potato. *Irrigation and Drainage, 50,* 129–137.
3. Epitome of Agriculture (2004). *Commissionerate of Agriculture,* Maharashtra State, Pune, India, 1–28.
4. Gundekar, H., & Khodke, U. (2009). Evaluation of crop coefficients for Rabi Sorghum (*Sorghum bicolarmonech*) in semi-arid climate. *Archives of Agronomy and Soil Science, 53*(6), 605–616.
5. International Crop Research Institute for Semi Arid Tropics (2004). Sorghum. Accessed on August 27, http://www.icrisat.org/sorghum/sorghum.htm.
6. Jadhav, S. S., Gutal, G. B., & Chaugule, A. A. (1990). Cost economics of drip irrigation system for tomato crop. In: *Proc. 11ᵗʰ International Congress on the Use of Plastics in Agriculture,* New Delhi, India.
7. Mohan, S., & Armugam, N. (1994). Crop coefficients of major crops in south India. *Agricultural Water Management, 26,* 67–80.
8. Narayanamoorthy, A. (2004). Drip irrigation in India: can it solve water scarcity. *Water Policy, 6*(2), 117–130.
9. Tyagi, N. K., Sharma, D. K., & Luthra, S. K. (2000). Evapotranspiration and crop coefficients of wheat and sorghum, *J. Irrig. Drain. Engineering, 126*(4), 215–222.

CHAPTER 11

WATER USE EFFICIENCY FOR MARIGOLD FLOWER (*TAGETES ERECTA* L.) UNDER FURROW AND DRIP IRRIGATION SYSTEMS

A. D. SIDDAPUR, B. S. POLISGOWDAR, M. NEMICHANDRAPPA, M. S. AYYANAGOWDER, U. SATISHKUMAR, and A. HUGAR

CONTENTS

11.1 INTRODUCTION

Drip/trickle or micro irrigation is a water saving method of applying the water near the roots of plants. It has been successfully used in almost all crops and in all types of climates including water deficit regions of the world. During 1991, India ranked seventh among 35 countries who used drip irrigation [1]. Reddy et al. [3] indicated that the area covered under drip irrigation is highest in Maharashtra followed by Andhra Pradesh and Karnataka. Soluble fertilizers and chemicals can be fertigated through drip irrigation in several dosages instead of 1–2 applications in conventional irrigation systems, thus increasing fertilizer use efficiency [1, 4].

This chapter presents research results to evaluate water use efficiency for marigold flower (*Tagetes erecta* L.) under furrow and drip irrigation systems.

11.2 MATERIALS AND METHODS

The experiment was conducted during October through March of 2012–2013 at Main Agricultural Research Station of UAS – Raichur (16°15' N latitude and 77°20' E longitude) with altitude of 389 m above mean sea level (MSL). The soil is clay loam in texture with a pH of 7.65. Seedlings of marigold (var. Orange double) were transplanted on 20 raised beds, which included 16 beds for drip irrigation and 4 beds for furrow irrigation. The 30 days old healthy and uniformly grown seedlings were used for transplanting. Transplanting was done in the morning hours with a spacing of 50 cm between rows and 45 cm between plants in drip treatments and in furrow treatment with spacing of 60 cm between rows and 45 cm between plants.

A light irrigation was given soon after transplanting. There were five irrigation treatments (60, 80, 100 and 120% of ET in drip irrigation and 120% of ET in surface irrigation) in randomized block design with four replications. Seedlings of marigold were transplanted at spacing of 50 cm x 45 cm. The seedlings were transplanted in 16 beds of 10 m x 0.8 m (drip) and 4 beds for furrow irrigation 10 m x 1 m. The 12 mm diameter lateral pipes were used in each bed with an inline dripper at 45 cm distance and discharge of 2 lph. Irrigation was provided daily after calculating water requirement based on past 24 hours of pan evaporation while in furrow irrigation it was scheduled once in seven days. The peak water requirement (Q) was calculated by:

$$Q = [A \times B \times C]/E \tag{1}$$

$$DI \text{ (Hours)} = \text{Dripper discharge}/(\text{Dripper spacing} \times \text{Inline spacing}) \tag{2}$$

$$E_a = (e \times q_{min} \times T/V) \times 100 \tag{3}$$

$$Ea = W_s/W_f \tag{4}$$

$$E_u = Y/WR \tag{5}$$

where, Q is the quantity of water required (mm day^{-1}), A is daily evapotranspiration (mm), B is canopy factor, C is crop coefficient, E is efficiency of drip irrigation system (percent), DI is duration of irrigation (hours), Ea is application efficiency (percent), e is total numbers of emitters q_{min} is minimum emitter flow rate (lph), T is total irrigation time (hours), V is total volume of water applied, (liters) [2], Ea is application efficiency of furrow irrigation, W_s is water stored in the root zone (liters), W_f is water delivered to the field, E_u is the water use efficiency of drip and furrow irrigation (kg m^{-3}), Y is crop yield (kg), and WR is total amount of water used in the field (m^3).

Amount of irrigation water applied to various treatments were based on daily pan evaporation readings. The irrigation treatments were imposed once the plants were established.

11.2.1 EXPERIMENTAL SETUP

The experimental set-up consisted of screen filter, mains, sub mains, laterals and emitters (drippers) and other accessories required for drip irrigation. The layout of drip irrigation is shown in Figure 11.1. General view of the experimental plot of marigold is presented in Figure 11.2.

11.3 RESULTS AND DISCUSSION

The first irrigation was applied up to field capacity in all the plots irrespective of different irrigation treatments. Subsequently, the irrigation water was delivered through drip irrigation as required in each treatment. In furrow irrigation, the crop was irrigated once in seven days and depth of irrigation was calculated. The amount of water applied per month for different levels of drip irrigation and furrow irrigation are presented in

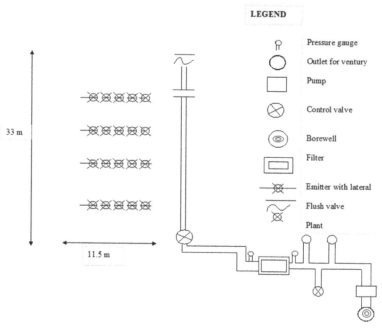

FIGURE 11.1   Layout of drip irrigation system and experimental set-up.

FIGURE 11.2   General view of the experimental plot of marigold.

Table 11.1. For drip irrigation at 60% ET, the monthly water requirement varied from 1.72 mm (October) to 58.82 mm (February), for 80% ET, water requirement varied from 2.29 mm (October) to 78.44 mm (February), for 100% ET water requirement varied from 2.86 mm (October) to 96.05 mm (February), and for 120% ET water requirement varied from 3.44 mm (October) to 115.33 mm (February); and for control plots (Furrow irrigation) the water requirement varied from 40 mm in October to 200 mm in December.

The water saving over furrow irrigation was maximum for 60% ET treatment (74.92%), followed by 80% ET (68.07%), 100% ET (61.45%) and 120% ET (54.65%) treatments. From these results it can be concluded that there was a substantial amount of water saving by drip irrigation system compared to furrow irrigation. This may be attributed to the fact that maximum amount of water applied will be stored in the root zone in case of drip irrigation treatments and the deep percolation losses are minimum. Further it could be observed that the water loss in furrow irrigation is more. These results are in agreement with the earlier findings of Tagar et al. [5].

Table 11.2 indicates that with increase in the level of irrigation the amount of water applied also showed an increasing trend, however irrigation capacity showed decreasing pattern. It was also observed that the irrigation capacity was lowest (1.13×10^{-4} ha m^{-3}) for furrow irrigation. The highest irrigation capacity of (4.53×10^{-4} ha m^{-3}) was obtained for the water application at 60% ET. Delta is the depth of irrigation (expressed in cm) required during the crop period. Delta of water for different treatments is presented in Table 11.2. It was observed from the table that delta was highest (88 cm) for furrow irrigation and among the drip irrigation treatments, it was lowest (22.07 cm) for water application at 60 per cent ET and it was highest (39.90 cm) for water application at 120 per cent ET.

The application efficiency for different treatments is given in Table 11.3. It is observed that application efficiency ranged from 92.48 (60% ET) to 88.49 (120% ET) for drip treatments and it was 82% for furrow irrigation treatment. This shows that the application efficiencies were higher in all the drip irrigation treatments as compared with the furrow irrigation treatment. It is seen that distribution efficiency ranged from 95.87 (60% ET) to 92.27 (120% ET) for drip irrigation treatments. For furrow irrigation the distribution efficiency was found to be 89.29% and it is less than

TABLE 11.1 Monthly Amount of Water Depth to Marigold Flower under Different Levels of Drip and Furrow Irrigation

Months	Amount of water applied through drip irrigation at different irrigation levels (mm)				Water applied in furrow irrigation (T5)
	T1 (60% ET)	T2 (80% ET)	T3 (100% ET)	T4 (120% ET)	
20th Oct	40	40	40	40	40
21st Oct 31st Oct	1.72	2.29	2.86	3.44	40
Nov	6.17	8.21	10.27	12.32	160
Dec	20.98	27.97	34.97	41.94	200
Jan	50.03	66.71	83.38	100.06	160
Feb	58.82	78.44	96.11	115.33	160
1st march to 18th March	42.98	57.30	71.63	85.96	120
Total	220.70	280.90	339.22	399.09	880
% saving water over furrow	74.92	68.07	61.45	54.65	

TABLE 11.2 Irrigation Capacity (duty) of 1 m3 of Water and Delta Value of Water for Different Treatments During the Crop Period

Treatments	Water applied in (liters plot^{-1})	Water applied in (m^3 ha^{-1})	Irrigation capacity (ha m^{-3})	Delta (cm)
T1	1765.6	2207.0	4.53×10^{-4}	22.07
T2	2247.2	2809.0	3.55×10^{-4}	28.09
T3	2713.7	3392.2	2.94×10^{-4}	33.92
T4	3192.4	3990.5	2.50×10^{-4}	39.90
T5	7040.00	8800.00	1.13×10^{-4}	88.00

that of all the drip irrigation treatments. The water use efficiency varied from (6.98 q ha^{-1} cm^{-1}, 3.62 q ha^{-1}cm^{-1}) for 80 and 120% ET under different drip irrigation treatments respectively as compared to 0.85 q ha^{-1}cm^{-1} in case of furrow irrigation method.

TABLE 11.3   Effects of Irrigation Methods and Irrigation Levels on Irrigation Efficiencies

Treatments	Application efficiency (%)	Distribution efficiency (%)	Field water use efficiency (kg m^{-3})
T$_1$	92.48	95.87	4.98
T$_2$	90.55	93.16	6.98
T$_3$	89.54	93.07	5.02
T$_4$	88.49	92.27	3.62
T$_5$	82.00	89.29	0.85

The application and distribution efficiencies were higher in all drip irrigation treatment than that of furrow irrigation. These findings are in agreement with earlier findings of other investigators [2, 6]. The higher application efficiency in drip irrigation compared to furrow irrigation system is due to the fact that, in drip irrigation we apply water as required by plant exactly and percolation losses below the crop root zone and the surface run off losses are very less, which results in more efficient application water.

The net returns and benefit–cost ratio for furrow and different drip irrigation levels are presented in Table 11.4. In this chapter, authors used 60.00 Rs. (Indian Rupees) = 1.00 US$. It is seen from the results that among all the drip irrigation treatments the highest net return of Rs. 2,44,300 per ha was obtained at 80% ET, closely followed by the treatment of 100% ET (Rs. 2,05,450 per ha) and the lowest net return was obtained in control treatment (Rs. 71,150.4 per ha). For 60% ET net return was found to be Rs. 1,15,150 per ha. Among all the drip irrigation treatments the lowest benefit–cost ratio of 1.73 was obtained in control treatment and the highest benefit–cost ratio was found in 80% ET (4.88) followed by 100% ET (4.10), 120% ET (3.33) and 60% ET treatment (2.3).

Projected additional returns from saved water are shown in Table 11.5. Using the saved water, highest additional yield of 41.82 tons could be obtained in treatment receiving water at 80% ET, When the water required to irrigate one ha marigold flower was completely used by drip irrigation,

TABLE 11.4   Economics of Furrow and Drip Irrigation Levels in Marigold Flower

Treatments	Crop yield (t ha^{-1})	Total returns (Rs ha^{-1})	Total cost of cultivation (Rs ha^{-1})	Net returns (Rs ha^{-1})	B:C ratio
T1	11.01	1,65,150	50,000	1,15,150	2.30
T2	19.62	2,94,300	50,000	2,44,300	4.88
T3	17.03	2,55,450	500,00	2,05,450	4.10
T4	14.46	2,16,900	50,000	1,66,900	3.33
T5	7.48	1,12,200	41049	71,150	1.73

In this chapter: 1.00 US$ = 60.00 Rs.; 1.00 q (quintal) = 100 kg.

the net returns were Rs. 9,21,600 per ha for 80% ET treatment and closely followed by the treatment receiving water 60% ET (Rs. 6,88,486). The lowest net returns were observed in drip irrigation treatments receiving water at 120% ET (Rs. 4,77,900) and (Rs. 1,12,200) per ha for furrow irrigation.

The extent of water saving under different methods of drip irrigation over furrow irrigation varies in ranges 74.92% (60% ET) to 43.62% (120% ET). But the additional yield with saved water was maximum in 80% ET level followed by 100% ET level. However the total yield was maximum in 80% ET level and which was 2.33 times more than furrow irrigation. Similar to these trends, the maximum projected net returns with drip irrigation were found in case of 80% ET level. These findings suggest the superiority of 80% ET level over other drip and furrow irrigation treatments. Thus it could be inferred that in sandy loam soils of semi-arid track, 80% ET level of drip irrigation can be ideal for successful marigold cultivation.

From the results of this investigation following conclusions have been drawn:

- The application and distribution efficiencies were found to be higher with drip irrigation treatments (T1 to T4) as compared to furrow irrigation.
- The water use efficiency was highest in 80% ET level (6.98 q ha^{-1}cm^{-1}) closely followed by 100% ET level (5.02 q ha^{-1}cm^{-1}) both under drip irrigation as compared to that in furrow irrigation (0.85 q ha^{-1}cm^{-1}), One q (quintal) = 100 kg.

TABLE 11.5 Projected Additional Return from Saved Water in Drip Irrigation Levels

Treatments	Water saved over furrow (%)	Yield (t ha⁻¹)	Net returns (Rs ha⁻¹)	Additional yield with saved water (t)	Total yield (t) (3+5)	Yield increase over furrow (t)	Increase in net returns with saved water (Rs)	Projected net returns with drip irrigation from water used in furrow (Rs) (4+8)
T1	74.92	11.01	1,65,150	32.88	43.89	36.41	9,93,200	6,58,350
T2	68.07	19.62	2,94,300	41.82	41.44	53.96	6,27,300	9,21,600
T3	61.45	17.03	2,65,450	27.14	44.17	36.99	4,23,036	6,88,486
T4	54.65	14.46	2,16,900	17.40	31.86	24.38	2,61,000	4,77,900
T5	—	7.48	1,12,200	—	—	—	—	1,12,200

In this chapter: 1.00 US$ = 60.00 Rs.; 1.00 q (quintal) = 100 kg.

- The highest yield (19.62 t ha^{-1}) was obtained for the treatment that received water at 80% ET level with drip irrigation which was closely followed by 100% ET level under drip irrigation (17.03 t ha^{-1}).
- The projected net returns were found to be highest in respect of 80% ET level (Rs. 9,21,600) followed by 100% ET (Rs. 6,88,486) if the saved water would be utilized for irrigation.
- In general considering the trend of different parameters, study suggests that water application at 80% ET using drip irrigation can be used to achieve higher flower yield with more returns compared to the water application with furrow irrigation system in marigold.

11.4   THE FUTURE LINE OF WORK

The present investigations suggest that Marigold flower crop responded well to drip irrigation. In the light of these findings further studies are required on the following aspects:
- Studies on effect of different drip irrigation levels on other varieties of marigold flower in different agro climatic situations.
- Studies on evaluation of marigold flower to different levels of fertigation.
- Studies on response of marigold flower to different types of water application devices such as micro and macro tube systems.

11.5   CONCLUSIONS

Marigold flower was grown on furrow and drip irrigation in year from October 2012 to March 2013 at New Orchard of main agricultural research station, UAS Raichur, India to evaluate comparative water use efficiency. Irrigation was applied though drip and furrow irrigation systems. Crop was planted on 50 cm x 45 cm under drip irrigation and 60 cm x 45 cm under furrow irrigation system. Water use efficiency was calculated as ratio of total yield (kg) to total water consumed by the crop (m^3).

Crop consumed less water under drip irrigation as compared to furrow irrigation system. The water use efficiency for marigold flower as influenced by irrigation methods and levels of drip irrigation. The water use efficiency was found to be 6.98 q ha^{-1}cm^{-1} and 3.62 q ha^{-1}cm^{-1} at 80 and 120% evapotranspiration (ET) under different drip irrigation treatments respectively as compared to 0.85 q ha^{-1}cm^{-1} in case of furrow irrigation method. Among different drip irrigation treatments, plant receiving water at 120% ET recorded the lowest water use efficiency (3.70 q ha^{-1}cm^{-1}) which went on increasing at 100 per cent ET (5.01 q ha^{-1}cm^{-1}) and 60% ET (4.98 q ha^{-1}cm^{-1}).

The value of water use efficiency at 80% ET was the highest (6.98 q ha^{-1}cm^{-1}) The water use efficiency in case of 80% ET level and 60% ET level were found to be significantly promising over all other treatments. All the drip irrigation treatments recorded higher benefit cost ratio of 2.30 to 4.88. The furrow irrigation gave a benefit cost ratio of 1.73. This suggested that drip irrigation has a greater scope for production of flower crops especially in water scarce areas.

11.6 SUMMARY

The present experiment was conducted with four drip irrigation levels and one furrow irrigation level. The objective of the study was to compare the yield and water use efficiency of the crop (marigold flower) under different levels of drip irrigations along with the conventional furrow irrigation.

It was observed that drip irrigation saved a lot of costly irrigation water than the furrow irrigation. Consequently, the water use efficiency of the crop was found to be higher in all the drip irrigated treatments than the furrow irrigation. The application and distribution efficiencies were found to be higher with drip irrigation treatments as compared to furrow irrigation. The highest yield (19.62 t ha^{-1}) was obtained for the treatment that received water at 80% ET level with drip irrigation Amongst the various drip irrigation schedules/drip irrigation levels, the crop irrigated by drip at 80% ET gave the highest water use efficiency (6.98 q ha^{-1}cm^{-1}) whereas the conventional furrow irrigation gave only 0.85 q ha^{-1}cm^{-1}. Considering the projected net returns with use of saved irrigation water, drip irrigation treatment with 80% ET is evaluated to give the highest return of Rs. 9,21,600.

KEYWORDS

- application efficiency
- benefit cost ratio
- canopy factor
- cost of cultivation
- crop coefficient
- delta
- distribution efficiency
- drip irrigation
- emitter
- evapotranspiration
- fertigation
- furrow irrigation
- irrigation
- marigold
- net return
- peak water requirement
- percolation
- water use efficiency
- yield

REFERENCES

1. Manjunatha, M. V. (2001). Micro irrigation need of the hour. *Kisan World*, November, 26–27.
2. Nakayama, F. S., & Bucks, D. A. (1986). *Trickle Irrigation for Crop Production, Design, Operation and Management. Elsevier Science Publisher*, The Netherlands, 376 pp.
3. Reddy, M., Ayyanagowder, M. S., Nemichandrappa, M., Balakrishnan, P., Patil, M. G., Polisgowdar, B. S., & Satishkumar, U. (2012). Techno economic feasibility of drip irrigation for onion (*Alluim cepa L*). *Karnataka J. Agric. Sci.*, *25*(4), 475–478.
4. Shock, C. C. (2001). Introduction of drip irrigation, Malheur Experiment Station – Oregon State University. *Information for Sustainable Agriculture*, 23–27.
5. Tagar, A., Chandio, F. A., Mari, I. A., & Wagan, B. (2012). Comparative study of drip and furrow irrigation methods at farmer's field in Umarkot. *World academy of Science, Engineering and Technology*, *69*, 863–867.
6. Wan-Shu Qin, K., Wang-Dan, L., & Feng-LiPing, L. (2007). Effects of saline water on cucumber yields and irrigation water use efficiency under drip irrigation. *Agric. Water Mgt.*, *23*(3), 30–35.

PART III

PRACTICES IN DRIP IRRIGATION DESIGN

CHAPTER 12

DEVELOPMENT OF SOFTWARE FOR MULTI CROP DRIP IRRIGATION DESIGN

A. BEHERA, CH. R. SUBUDHI, and B. PANIGRAHI

CONTENTS

12.1 INTRODUCTION

Water is a main source for living. Water contributes a vital input in food production in form of irrigation. To achieve required food production with increasing population, India has an irrigation potential of 139.89 million

This chapter is an edited version: Anuradha Behera, "Development of a software for multi crop drip design." 2014. Unpublished thesis for Master of Technology, Department of Soil and Water Conservation Engineering, College of Agricultural Engineering and Technology, Orissa University of Agriculture and Technology, Bhubaneswar, Odisha, India hectares [11, 13]. Due to increase demand for water in various sectors such as domestic, industry, agriculture, hence the country will face water stress in the coming years. Therefore hand in hand with technologies for water harvesting and storage, technologies for precision water application methods need to be adopted. Micro-irrigation technology is the most efficient irrigation method available as at present. Micro-irrigation mainly deals with drip and sprinkler irrigation. Drip irrigation is becoming popular day-by-day due to efficient use of water and increased productivity.

Drip irrigation also called trickle irrigation is a pressurized system to irrigate the crops and orchards, consists of an extensive network of pipes usually of small diameters that deliver water directly to the soil near the plant. It is the slow application of water directly to the plant's root zone. Drip irrigation can provide optimum moisture level in the soil at all times, resulting in less water lost to the sun and wind. In drip irrigation, water is not wasted, as the root zone area is maintained at a steady moisture level, combining the proper balance of water and air for a very efficient irrigation system. Drip irrigation can be used to ensure proper watering of all of plant needs. Drip irrigation can be applied to all types of crops grown in rows. Filters are used to remove suspended materials, organic matter, inorganic impurities to reduce blockage of the drippers and better functioning of valves. At pumping station, control valves and pressure valves are installed to provide required pressure heads to the system.

Drip irrigation system supplies water as per the need of the plant and maintains optimum soil moisture at root zone level. It saves water up to 30–70% as compared to surface irrigation and increases yield as well. The crop cultivated under drip irrigation system that gives better yield and of good quality with high return per unit area. It saves labor cost as less money is spent on weeding and intercultural practices. It minimizes soil erosion. It saves nutrients, as the application of fertilizer through irrigation water is possible thereby, saving in fertilizer and improved fertilizer use efficiency. Poor quality water can also be used more safely. It can be irrigated in undulated land. Better pest and disease management can be done through drip irrigation system. Drip irrigation is an eco-friendly technology. It has lower operating pressure comparable to sprinkler irrigation means reduced energy costs for pumping. Drip irrigation gives highly uniform distribution of water.

Drip irrigation is adopted to save water, energy and manpower in agriculture. The system can either be operated manually or automatically. Widespread application of drip irrigation is highly accepted by the farmers. The success of drip irrigation depends on the proper design of the system. As the design of drip irrigation is complicated and time consuming, hence simplified drip irrigation model need to be develop for efficient design of drip irrigation system with less time. Design of drip irrigation for fruit crops can be done easily with use of a model with little time for different area and different sizes of orchard.

The present study presents user-friendly software for design of drip irrigation system, with the following objectives:

- To compute crop water requirement of different crops using software which can be irrigated by drip irrigation system.
- To develop a user-friendly software for design of drip irrigation system.

12.2   REVIEW OF LITERATURE

12.2.1   DRIP IRRIGATION

Drip or trickle irrigation refers to the frequent application of small quantities of water at low flow rates and pressures. Rather than irrigating the entire field surface, as with sprinklers, drip irrigation is capable of delivering water precisely at the plant where nearly all of the water can be used for plant growth. The uniformity of application is not affected by wind because the water is applied at or below the ground surface. A well-designed and maintained drip irrigation system is capable of an application efficiency of 90% [14].

Drip irrigation is one of the latest methods of irrigation, which is becoming increasingly popular in areas having water scarcity and moderate salt problems. In drip irrigation, water is directly applied to the root zone of plant and it permits the irrigator to apply the volume of water closely matching the consumptive use of plant. Design of drip irrigation system depends on several parameters including topography, soil type, crop to be irrigated, weather conditions, technological and financial resources. Different criteria are available for designing the drip irrigation system for widely spaced row crops such as orchard and vegetables for supplying the water to individual plants with the help of a single or a set of dripper based on their rooting pattern and canopy area. In this situation, there is no need to apply water to the entire land area and lateral are generally spaced along the plant rows. For closely spaced field crops, the entire land area needs to be wetted and the drip irrigation system needs to be designed on the basis of meeting the water requirement of the total cultivated area [3].

12.2.2 *DESIGN PARAMETERS OF DRIP IRRIGATION SYSTEM*

The procedure for the design of drip irrigation system is classified into [6]:
- Crop water requirement.
- Major components for design drip irrigation system.
- Layout of the irrigation system.
- Irrigation time requirement.

12.2.2.1 Crop Water Requirement

The estimation of the water requirement of crop is one of the basic needs for crop planning on the farm and for planning of any irrigation. Water requirement may be defined as the quantity of water, regardless of its source, required by a crop or diversified pattern of crops in a given period of time for its normal growth under field condition. Water requirement includes the losses due to evapotranspiration or consumptive use plus the losses during the application of water (unavoidable losses) and the quantity of water required for special operations like land preparation, pre-sowing irrigation and transplanting [7].

Determination of crop water requirement largely depends upon availability of climatic data records. Water requirement can be calculated by the following formula [3]:

$$WR = (A \times B \times C \times D)/E \tag{1}$$

where, WR = Crop water requirement for the growing period, mm/day; A = Potential evapotranspiration, mm/day; B = Crop factor (depends on growth stage and foliage cover); C = Canopy factor [= Area of plant shadow at noon ÷ (plant spacing x row spacing)]; D = Crop area, (row-to-row spacing in m crop-to-crop spacing in m), m²; and E = Emission uniformity of drip irrigation system, decimal.

12.2.2.2   Major Components for Design of Drip Irrigation System [6]

Pumping unit
It is the unit which supplies water from water source. It conveys water to the whole irrigation system. Source of water and water quality should be check.

Filtration system
A filter unit removes the suspended impurities in the irrigation water, so as to prevent blockage of holes and passage of drip nozzle. The type of filtration needed depends on water quality and emitter type. There are various types of filtration unit such as hydro-cyclone or centrifugal sand separator, sand and gravel media filters, screen filter, disc filter.

Main line
The mainline conveys the water from the filtration system to submain. It is normally made of rigid PVC. The diameter of the pipeline based on flow capacity. Pipeline of 40 to 110 mm diameter and above with a pressure rating 4 to 6 kg/cm² are used for main pipes.

Submains
The submain conveys the water from mainline to laterals. It is made of rigid PVC material. The diameter ranging from 32 mm to 75 mm having pressure rating of 4 to 6 kg/cm² are used.

Laterals
Laterals are small diameter flexible pipes or tubing made of low-density polyethylene (LDPE) or linear low-density polyethylene (LLDPE) and having 12 mm, 16 mm and 20 mm diameter. They are colored black to avoid algae growth inside and minimize the damaging effect of ultra-violet radiation. They can withstand maximum pressure of 4 kg/cm².

Emitters/drippers

They discharge water from lateral onto the soil near the plants. Emitters are available with range of flow between 2 to 8 liters/hour. They are made of plastic such as polyethylene or polypropylene. The dripper can be classified according to working principle, discharge, type, structure, working pressure, durability, pressure compensating and non-pressure compensating drippers. The main principle in dripper selection is to achieve the minimum discharge with maximum size of water passage.

Valves and accessories

- *Control valves* are used to control the flow through the sub-main pipes. They are installed on filtration system, mainline and on all sub-mains. They are made up of gun metal, PVC, cast iron and the size ranges from 20 mm to more than 140 mm.
- *Flush valve* is provided at the end of each sub-main to flush out the water and dirt accumulated at the end of sub-main.
- *Air release cum vacuum breaker valve* is provided at the highest point in the mainline to release the entrapped air during the start of system and to break vacuum during shut off. It also provided on sub-main if sub-main length is more.
- *Non return valve* is used to prevent the damage of pump from backflow of water hammer in rising mainline of drip irrigation system.
- *Pressure and flow regulators* required in the control head to ensure instant flow into the drip irrigation system.
- *End cap* are used to close the lateral ends, sub-main ends or mainline ends. Submain and mains are preferably provided with flush valve. They are convenient for flushing the line.

12.2.2.3   Layout of Drip Irrigation System

Layout of a typical drip irrigation system is shown in Figure 12.1.

12.2.2.4   Irrigation Time

It is an important aspect of determining the time to irrigate and how much is to be applied (irrigation depth) in each irrigation. Proper irrigation is

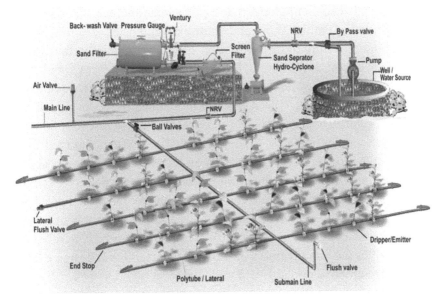

FIGURE 12.1   Typical layout of drip irrigation system. [Courtesy of Jain Irrigating System Limited: www.jains.com]

essential for the efficient use of water and other inputs in crop production. It is planned to either fully or partially provide the estimated water requirement of the crop [7]. Irrigation time is the ratio of water requirement to the discharge of all the drippers per plant. It is the time required to irrigate the land [6].

12.2.3   DESIGN OF DRIP IRRIGATION SYSTEM

Simple computerized drip irrigation design was designed by Feng et al. [5]. They found that uniformity of a drip irrigation system is affected mainly by hydraulic design, manufacturer's variation, temperature effects, and plugging. A simple approach using the energy gradient concept and mean discharge approximation was employed in developing a computer-aided design program for drip irrigation for both a single lateral and a submain unit. The calculation technique, by revising the energy gradient line of lateral and submain, is defined as the revised energy gradient line (REGL) approach. The REGL approach was tested and verified by the step-by-step calculations, which is most accurate, but is time consuming and impractical. The computerized design also can include the manufacturer's variation of emitter flows and emitter plugging.

Emission uniformity of drip irrigation system is very important while designing the system and it should be as high as possible [9]. A simulation model for design and evaluation of micro-irrigation systems was developed by Pedras et al. [9]. The AVALOC model has been developed for design and performance analysis of micro-irrigation systems, adopting the sector as the unit for analysis. Model computations were supported by a database containing updated information on the emitters and pipes available on the market, and where the information relative to the sectors being designed or evaluated was stored. In the design mode, the model provides for the selection of pipes and emitters that permit the attainment of the target performance, including emitter discharge uniformity. The research paper describes the main features of the model and shows a design example applied to an olive orchard.

Investment decision model for drip irrigation system was carried by Srivastava et al. [12]. This software program has been developed for estimating the threshold economic value of the investment cost of drip irrigation. In addition to the threshold value of investment cost, the software provides information on energy consumption and net return. The software can be used both for annual crops such as sugarcane or seasonal crops such as vegetable rotations (winter–summer). To demonstrate the interdependence of various input parameters, an analysis had been made using local data in this software. The analysis had provided the relationship between the investment cost and the yield gain factor, the returns from the crop, as well as the savings in energy and the size of the prime mover with regard to the size of the farm. The output shows that if the actual investment cost of the drip irrigation is below this level, it will be economical. The output also gives energy saving per annum and saving in the size of the prime mover.

A software tool to display approximate wetting patterns from drippers with advances in system design and management named as WetUp was developed by Cook et al. [4]. This user-friendly software tool, WetUp, has been developed to highlight the impact of soils on water distribution in trickle-irrigated systems. Wet- Up determines the approximate radial and vertical wetting distances from an emitter in homogeneous soils calculated using analytical methods. On the results and discussion, these wetted patterns are relatively spherical (for the buried emitter) or semi-spherical (for the surface emitter) in shape due to capillary forces dominating the flow and concluded that, WetUp gives a reasonable estimation of the wetted perimeter arising from infiltration from both buried and surface point sources & provides visualization of wetting patterns and how changing the soil properties, position of the emitter, or the volume of water applied will change the wetting patterns in homogeneous soils.

Aa rapid manufacturing method for water-saving emitters for crop irrigation based on rapid prototyping and manufacturing was developed by Zhengying et al. [16]. In this research work, a rapid prototyping and manufacturing (RP&M) technique was used to develop a new type of emitter and its mold. A rapid manufacture technique for water-saving drip irrigation emitters, based on RP&M technology, was studied. The CAD design of an emitter using parameterized design, CAD process design and the generation of a rapid tooling process model of an emitter were described. Prototypes had been built using the rapid tooling technique to complete the rapid verification and the modification of the emitter design, and with the manufacture of a precision mold as the basis, the design and manufacturing of the emitter using a metal spraying mold has been carried out.

Software namely DRIPD was developed by Rajput et al. [10] at Indian Agricultural Research Institute, New Delhi. It gives the brief idea of design of drip irrigation system. Yurdem et al. [15] worked on development of a mathematical model to predict head losses from disc filters in drip irrigation systems using dimensional analysis. A model was developed using dimensional analysis to predict head losses in disc filters. Three different filter designs, each with four different inlet and outlet pipe diameters, were used to measure head losses at different flow rates in the laboratory. The parameters influencing head losses were considered to be the inside diameters of the inlet and outlet pipes, the inside diameter of the filter body, the inflow and outflow area where the inlet and the outlet pipes intersect with the body of the filter, the, effective length of filter disc group, the outside and inside diameter of the filter disc, the water velocity in the inlet pipe and the kinematic viscosity of water. A dimensional analysis was carried out using Buckingham's pi-theorem. To develop the model, experimental head loss data from 12 filters were considered. The model accounted for 90.18% of the variation in the pressure coefficient.

Pocket PC software to evaluate drip irrigation lateral diameters with on-line emitters was developed by Molina-Martíneza et al. [8]. It was developed with a LabVIEW PDA® as programming language and enables engineers and installers to calculate commercial diameters to be used in laterals of drip irrigation, without the need for being at the personal computer. Specifically, this software allows users to immediately evaluate the sensibility to changing demands (e.g. crop, water needs, spacing, etc.) in all the range of commercial diameters for drip lines with on-line drippers. Input data required were drippers flow rates, number of drippers, spacing between the drippers, medium pressure in the lateral and pressure tolerance. As results, it shows with a figure with light emitting diodes, the commercial diameters that can be used. Other results implemented in this tool were the maximum and minimum pressures for each diameter and a table that shows, on a comparative basis, whether the pressure tolerance of every diameter is exceeded or not. These calculations were executed at once for every commercial diameter in a pocket PC. This allows to users makes these calculations immediately in the field without the need of moving to the office and using a PC.

Work on development of software design of drip irrigation system has also been done by other researchers [3]. The software was developed to design of drip irrigation system, with taking onto calculation of crop water requirement, friction head loss main line and operating pressure of system, which are very important for drip irrigation system. Minimum inputs were provided to software and tested to size of lateral, sub mains main lines with their lengths and friction loss.

Arbat [1] developed a model namely Drip-Irriwater: computer software to simulate soil-wetting patterns under surface drip irrigation. The shape and dimensions of the volume of wet soil below the emitter were some of the most influential variables in the optimal design and management of drip irrigation systems. This paper presents Drip-Irriwater, a code for determining soil wetting patterns under a single emitter in order to suggest planning factors for a drip irrigation system. The code solved Richards' equation using the finite difference method subject to appropriate boundary conditions for drip irrigation. The boundary conditions on the surface of the soil allow us to consider forming a pond under the emitter. The user interface enables users to simply enter the input parameters (discharge rate, soil type and horizons, irrigation time, initial water content and total simulation time). The code then displays the soil water content and pressure heads, enabling a visual distinction between the wetted radius and depth, the parameters required for subsequent drip irrigation design. The results obtained with Drip-Irriwater were compared with those obtained with HYDRUS and with results from field tests carried out on three different soil types. The soil water distribution, as well as the wetted radius and depth, calculated from the Drip-Irriwater and HYDRUS code, were very similar.

Simple design software for drip irrigation system has been developed by Behera et al. [2]. In this work, authors developed software for design of drip irrigation system for Sapota crop. Here inputs for the software are crop type, land size, climatologically parameters where the software gives the output total peak water requirement, depth of irrigation applied, design of dripper, selection and design of dripper, lateral, submain and mainline, power requirement of the pump and irrigation time. The limitation of the software is it only described about only one crop specification that is Sapota.

12.3 MATERIALS AND METHODS

Drip irrigation is the modern irrigation system, which is associated with very costly equipments. Hence before installation of the system, it is necessary that planning should be done with due care. In this project work the drip irrigation software has been developed for drip design.

12.3.1 BASIC INFORMATION REQUIRED FOR DESIGN

- **Survey of land:** Primarily the land is to be surveyed whether plain or undulating, if it is not leveled then the land should be leveled. Then the proper design and layout for the land is selected.
- **Type of crop:** It is required for the lateral and emitter spacing, irrigation water requirement because different crops having different row-to-row and crop-to-crop spacing which is given in Annexure I.

- **Soil type:** Soil characteristics is one of the main part of the irrigation design because of texture and structure of soil, as it required for soil infiltration rate, water holding capacity, bulk density, spacing of dripper, spacing of emitter and scheduling of irrigation, irrigation interval and spacing.
- **Land slope:** Determine lateral, main and sub-main diameter with the consideration of frictional head loss.
- **Climatic record:** This gives the ideas about the crop water demand, irrigation scheduling of the crops.
- **Source of water and water quality:** It is required for the selection of filter and quality of water, source of water may be ground water, tube well, well or pond.

After the collection of preliminary information, mathematical steps are to be followed for the design of components used in drip irrigation system. The design of drip irrigation system is mainly associated with the calculation of peak water requirement, type of dripper selected which is suitable for the irrigation, number of plant in the field, dripper required for each tree, length of the lateral line, number of plants should present in one lateral, total number of dripper used in the field, discharge rate of each lateral, submain line, main line, appropriate inside and outside diameter of laterals, submain line, mainline, length of laterals, submain line, mainline, number of laterals on submain and horse power requirement of the pump, which will be efficient to supply the required amount of water and irrigation time. These factors are important part of the drip irrigation design which has been solved in the steps by some mathematical formulas.

The design steps are as followed, where the primary data are the area of the land, length on the land, width of the land, crop to be irrigated, row-to-row and crop-to-crop spacing. With these data and the followed formulae will present the whole design of the drip irrigation system.

12.3.2 DESIGN STEPS FOR DRIP IRRIGATION SYSTEM

12.3.2.1 Step I: Determination of Crop Water Requirement

Peak water requirement (l/day/plant) is given below [6]:

$$q = (A \times PE \times Pc \times Kc \times w)/E_u \qquad (2)$$

where, A = area of the crop (row-to-row, m × plant-to-plant, m), m²; PE = maximum pan evaporation, mm/day; Pc = Pan Coefficient, approximately taken as 0.7 to 0.8; Kc = crop coefficient; w = wetted area of the crop, %; and E_u = emission uniformity of drip system, decimal.

Crop coefficients for any crop depend upon foliage characteristics, stage of growth of crop, environment and geographical location. It is the area which is shaded due to its canopy cover when sun is over head, which depends upon stage of growth of plant. Percentage wetted area covered of crop varies from 1/3 to 2/3 of area of the plant.

12.3.2.2   Step II: Selection of Dripper

Number of dripper or emitter required per tree is given as:

$$n = (w \times A)/\text{Effective wetted area by one Emitter} \tag{3}$$

Total number of plant,
$$T_p = A_1/A \tag{4}$$
Total no. of dripper,

$$N = T_p \times n \tag{5}$$

where, A_1 = Area of the land, m².

After the design of dripper, the layout is to be selected. According to the layout we can decide about the design and selection of lateral, submain and mainline. Generally two types of layout are found to irrigate with the drip irrigation system. Figures 12.2 and 12.3 show the views of two types of layout systems.

For Layout 1, the lateral passes throughout the width of the land and submain along the length of the land. For layout 2, the submain crosses on the middle of the field hence the lateral is divided into two sections. Length of the lateral is just half of the length of field.

12.3.2.3   Step III: Selection and Design of Lateral

Length of lateral depends on the layout of the drip irrigation system.

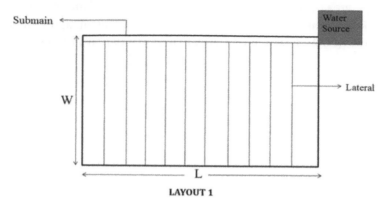

FIGURE 12.2   Layout 1 for drip irrigation system.

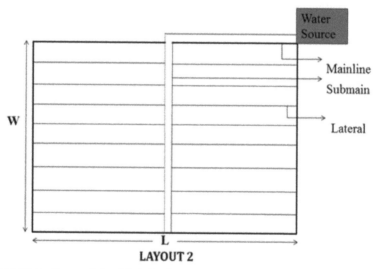

FIGURE 12.3   Layout 2 for drip irrigation system.

For layout 1, Length of Lateral $L_{lateral1} = W$ (6)

For layout 2, Length of Lateral, $L_{lateral2} = L/2$ (7)
No. of plant on one side of the lateral $= L_{lateral}/cc$ (8)

Total no. of dripper on one lateral = No. of plant in one lateral
                                                        × Number of dripper per plant   (9)

Discharge rate of one lateral:

Q_l = Number of dripper in one lateral
          × Discharge through each dripper   (10)

where, W = width of the land, m; L = Length of the site, m; $L_{lateral}$ = length of lateral, m; and cc = crop-to-crop spacing, m.

Diameter of the lateral is selected from the available sizes of pipe diameter of 12 mm, 16 mm, and 20 mm. Considering the thickness of the pipeline, the inside diameter of the pipeline is selected. Then the head loss due to friction is calculated. The friction loss for pipeline can be computed from the Hazen-Williams equation. Head Loss in lateral [6] is:

$$\Delta H_{lateral} = (K \times Q_{lateral}/C)^{1.75} \times D_{lateral}^{-4.871} \times L_{lateral} \times F \qquad (11)$$

where, $\Delta H_{lateral}$ = friction loss in lateral, m; K = a constant, 1.21×10^{10}; $Q_{lateral}$ = flow rate in the lateral, lps; c = friction coefficient for continuous section of pipe; $D_{lateral}$ = inside diameter of the lateral, mm; $L_{lateral}$ = length of the lateral, m; and F = outlet factor.

12.3.2.4   Step IV: Selection and Design of Submain

According to the design adopted for the drip irrigation, submain passes through the middle of the field.

For layout 1, Length of the submain ($L_{submain}$) = Length of the land (L)
For layout 2, Length of submain ($L_{submain}$) = Width of the land (W)
Number of laterals on one side of the submain

$$N_l = L_{sub\ main}/(rr) \qquad (12)$$

For layout 2, total number of lateral on submain

N_l = No of lateral on one side of submain × 2   (13)

No of lateral on one side of submain = $L_{sub\ main}/(rr)$ (14)

Discharge rate of submain, liters/sec

$$Q_{\text{sub main}} = Q_{\text{lateral}} \times N_1 \qquad (15)$$

Head Loss in lateral:

$$\Delta H_{\text{submain}} = K \times (Q_{\text{submain}}/C)^{1.75} \times D_{\text{submain}}^{-4.871} \times L_{\text{submain}} \times F \qquad (16)$$

where, Q_{lateral} = discharge rate of one lateral, lps; N_1 = total number of lateral on sub main; L_{submain} = length of submain, m; rr = row-to-row spacing, m; = friction loss in submain, m; K = a constant, 1.21 10^{10}; Q_{submain} = flow rate in the sub main, lps = friction coefficient for continuous section of pipe; D_{submain} = inside diameter of the submain, mm; L_{submain} = length of the submain, m; F = outlet factor.

12.3.2.5   Step V: Selection and Design of Main Line

For Layout 1, Length of mainline, L_{main} = Distance from water source

For Layout 2, Length of mainline, L_{main} = Length of the land/2

Discharge rate of mainline is approximately equal to the discharge through all submains. The friction loss for pipeline can be computed from the Hazen-Williams equation. Head Loss in main:

$$\Delta H_{\text{main}} = K \times (Q_{\text{main}}/C)^{1.75} \times D_{\text{main}}^{-4.871} \times L_{\text{main}} \qquad (17)$$

where, ΔH_{main} = friction loss in main, m; K = a constant, 1.2110^{10}; Q_{main} = the flow rate in the main, lps; c = friction coefficient for continuous section of pipe; D_{main} = inside diameter of the main, mm; and L_{main} = length of the main, m.

12.3.2.6   Step VI: Selection of Pump

Total pressure head drop in meters due to friction:

$$H_f = \Delta H_{\text{lateral}} + \Delta H_{\text{sub main}} + \Delta H_{\text{main}} \qquad (18)$$

Total head in m:

$$H = H_s + H_d + h + H_f + H_1 + D \qquad (19)$$

where, $\Delta H_{lateral}$ = head loss in lateral, m; $\Delta H_{sub\ main}$ = head loss in submain, m; = Head loss in main, m; H_s = suction head, m; H_d = delivery head, m; h = operating pressure head of drip system, m; H_f = total pressure head drop due to friction, m; H_1 = filter loss + fitting loss + venturi loss, m; and D = elevation difference in m.

Power requirement of pump is given as:

$$Hp = (Q \times H)/(75 \times \eta_{motor} \times \eta_{pump}) \qquad (20)$$

where, Hp = power of the pump, hp; H = total head, m; η_{motor} = motor efficiency, %; and η_{pump} = pump efficiency, %.

12.3.3 IRRIGATION TIME

Irrigation time or duration (hours) is the ratio of crop water requirement to the discharge of all the drippers per plant. It is calculated by the formula:

Irrigation Time = Crop Water Requirement/(Discharge of dripper × No.of dripper per plant)   (21)

Therefore, the design of drip irrigation system is a collection and solution of several of mathematical equations (Eqs. (1)–(21)). The design of drip irrigation is associated with the primary consideration like crop type, land size, crop water requirement and other weather parameters. Then the component of the drip irrigation system is taken into consideration. It is mainly associated with the design of dripper, lateral, submain and main line and pump to irrigate. Other components like filter, valves and fertigation system are also needed to be designed. After designing the drip irrigation components, the irrigation time is to be calculated.

12.3.4 SOFTWARE DEVELOPMENT

After studying the design of drip irrigation system, the software was developed by using Visual Basic 6.0. For easy calculations of the mathematical equations, those are used in drip irrigation design to develop the software. The developed software program has advantages that the design can be adopted for all type of fruit crops, but in this chapter the software was evaluated only for three crops: lemon, sapota and coconut. After collection of necessary data like crop type, crop spacing, area of the land, soil characteristics, the values are used in the software to get the design of the drip irrigation system. The software showed that the design of drip irrigation system is easy for a general farmer to understand. It avoids lot of calculations. As drip design is complicated and installation cost is too high, the software can give us a precise, simple, economic and easy design. The flow chart is given in Figure 12.4 for the design steps that are a preliminary stage to develop the software.

12.3.4.1   About Visual Basic

Visual Basic is a third-generation event-driven programming language and integrated development environment (IDE) from Microsoft released in 1991. VISUAL BASIC is a high level programming language which evolved from the earlier DOS version called BASIC. BASIC means Beginners All-purpose Symbolic Instruction Code. It is a very easy programming language to learn. The code looks like English Language. Different software companies produced different versions of BASIC, such as Microsoft QBASIC, QUICKBASIC, GWBASIC, IBM-BASICA and so on. However, people prefer to use Microsoft Visual Basic today, as it is well developed programming language and supporting resources are available everywhere. Now, there are many versions of VB exist in the market, the most popular one and still widely used by many VB programmers is none other than Visual Basic 6. We also have VB.net, VB2005, VB2008 and the latest VB2010. Both Vb2008 and VB2010 are fully object oriented programming (OOP) language.

12.3.4.2   Visual Basic Scripts

The Visual Basic 6 integrated programming environment is shown in Figure 12.5. It consists of the toolbox, the form, the project explorer and the

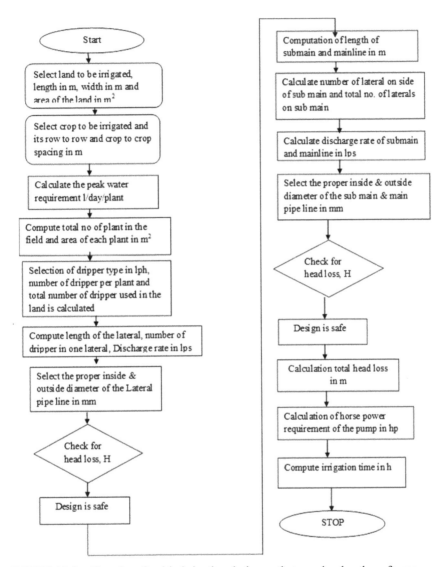

FIGURE 12.4   Flowchart for drip irrigation design so that one develop the software.

properties window. Form is the primary building block of a Visual Basic 6 application. A Visual Basic 6 application can actually comprises many forms; but we shall focus on developing an application with one form first. The source code window for Form-1 as shown in Figure 12.6 will

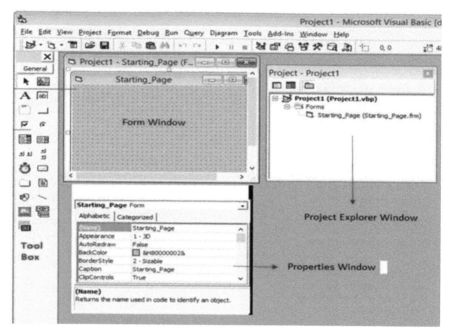

FIGURE 12.5   VB6 programming environment.

appear. The top of the source code window consists of a list of objects and their associated events or procedures. In Figure 12.6, the object displayed is Form and the associated procedure is Load. Each object has its own set of procedures. One can always select an object and write codes for any of its procedure in order to perform certain tasks. The beginning and end statement of a programming is given below: Private Sub <selected object>End Sub.

12.3.4.3 Example Program

In this program, two text boxes are inserted into the form together with a few labels. The two text boxes are used to accept inputs from the user and one of the labels will be used to display the sum of two numbers that are entered into the two text boxes. Besides, a command button is also programmed to calculate the sum of the two numbers using the plus operator.

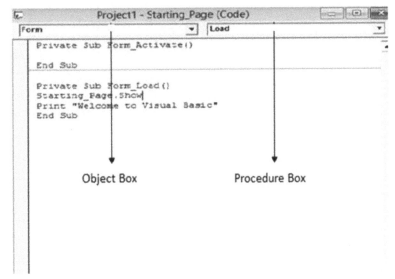

FIGURE 12.6   Source code window.

The program use creates a variable sum to accept the summation of values from text box 1 and text box 2. The procedure to calculate and to display the output on the label is shown below. The output is shown in Figure 12.7.

Private Sub Command1_Click()
Sum = Val(Text1.Text) + Val(Text2.Text) (Codal Procedure)
{To add the values in text box 1 and text box 2} (Explanation)
Label1.Caption = Sum (Codal Procedure)
{To display the answer on label 1} (Explanation)
End Sub

12.3.5   DESIGN PARAMETERS FOR SOFTWARE DEVELOPMENT

The design of a drip Irrigation system involves estimation of the following parameters.

- Area to be irrigated, length and width of the land in m.
- Type of crop to be irrigated (for the development of software the crop taken to consideration are lemon, sapota and coconut), crop

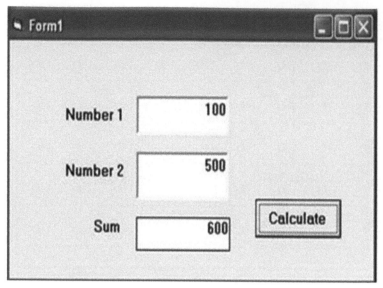

FIGURE 12.7   Output box.

spacing and area of plants. The required data like crop spacing and crop coefficient are required to design the system.

- Peak water requirement of a plant per day and estimation of total water requirement for a given area.
- Selection of dripper type (2 lph, 4 lph or 8 lph), number of dripper per plant and amount of water discharge per hour through each emitter.
- Layout of the system considering topography, field shape and location of the water source.
- Selection and design of lateral, submain and main line. Length, diameter and discharge rate of the pipelines.
- Calculation of head loss of the pipe lines like lateral, submain and main line. Check for head loss.
- Selection of filters and other equipment.
- Water required to be pumped from the well. Horse power of a pump set. This depends on discharge and the total head including friction losses over which water is to be lifted or pumped.

The flow chart for design of drip irrigation system is shown in Figure 12.4. At first, designer needs to select the land, its area including length and width. Then he has to select the crop for which drip system is to be designed. Various crops have different crop geometries. So for the particular crop, he has to find out the spacing from row-to-row and plant-to-plant. Peak water requirement of the crop has to be calculated. Discharge of dripper and total number of drippers used in the area have to be computed so that he can calculate total water requirement of the field during peak period. Irrigation timing has to be estimated depending on the dripper discharge and volume of water requirement of the crop. Then he has to compute the head losses in the system and also length of laterals, submains and mains in the system. Finally the power requirement of the pump has to be computed by knowing the suction and delivery heads, head losses in the system and total discharge carried in the main pipe.

12.4 RESULTS AND DISCUSSION

12.4.1 *GRAPHICAL USER INTERFACE OF MULTI-CROP DRIP IRRIGATION DESIGN*

One of the main objectives of the project is to develop user-friendly software for drip design. To start with, the screen shot below (Figure 12.8) shows the welcome screen to the user interface.

The welcome screen directs the user to main design input parameter screen (Figure 12.9) which will be used insert values of all parameters. The main design parameters affecting the drip irrigation system include: land area, type of crop grown, crop spacing, climatic condition like maximum pan evaporation and pan coefficient (0.7 to 0.8), crop coefficient of the crop, percent wetted area. These parameters are required for the software to calculate depth of irrigation and peak water requirement. Then there is 'NEXT' button to proceed for the next page.

From the previous screen shot, the user will obtain the peak water requirement based on the input parameters entered. Once the peak water requirement is given, this will be followed by next step which includes design of dripper. The design of dripper will be based on the peak water requirement and hence the graphical user interface gives for the selection of dripper. There are three different values of dripper type: 2, 4 and 8 lph. The user may choose the nearest value of the type of dripper values lies

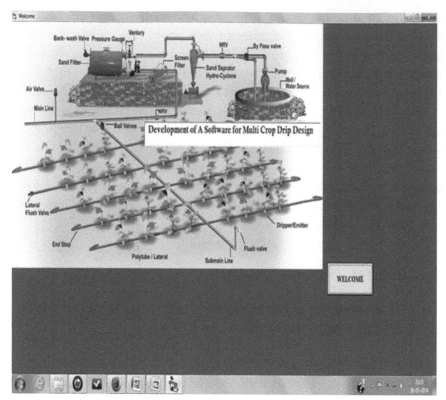

FIGURE 12.8   Welcome screen for drip design model , a modified version of Fig. 12.1 in this chapter.

in the between the given values. Once the dripper is chosen, the user is instructed to click on the respective boxes to obtain number of dripper plant, total no. of drippers that to be used in the land and total no. of plants in the land to be planted. Then there is a 'NEXT' button to proceed and a 'BACK' button. If user wants to correct any value entered in the previous page, then 'BACK' button is to be clicked and verified. Figure 12.9 shows the window box of design of dripper system.

After developing the dripper design, the user has to choose the layout of drip irrigation system to be placed in agricultural field. The layout types have already been shown in Figures 12.2 and 12.3. The below screen shot (Figure 12.10) shows the different layouts that can be chosen by user. Particular layout can be chosen by clicking on layout name. Then there is a 'NEXT' button to proceed and a 'BACK' button. If user wants to correct

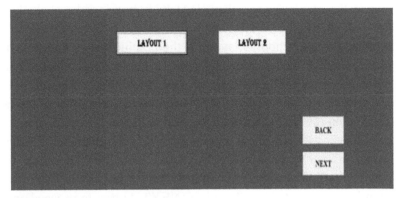

FIGURE 12.9 Design input parameter screen.

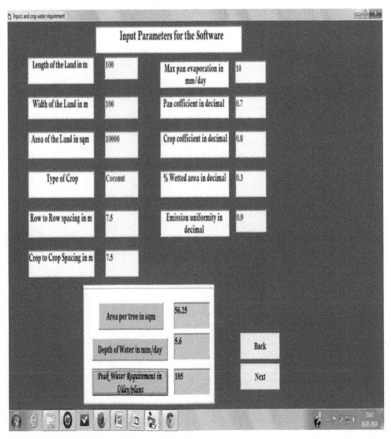

FIGURE 12.10 Selection of layout.

any value in the previous page, then 'BACK' button is to be clicked and verified.

Assuming the user chooses layout 1, the graphical user interface has been shown in the figure below (Figure 12.11). Lateral pipelines are designed in this screen. The software provides the output for length of the lateral, discharge rate of the lateral. It also computes the number of laterals that to be laid on the land to irrigate. According to the discharge, diameter of the pipeline is to be selected. After selection of the diameter of lateral head loss calculated by the software. Head loss calculation determines that the pipeline is sufficient enough to carry the flow or not, which determines the design is safe or not. Then there is a 'NEXT' button to proceed and a 'BACK' button. If user wants to correct any value in the previous page, then 'BACK' button is to be clicked and verified.

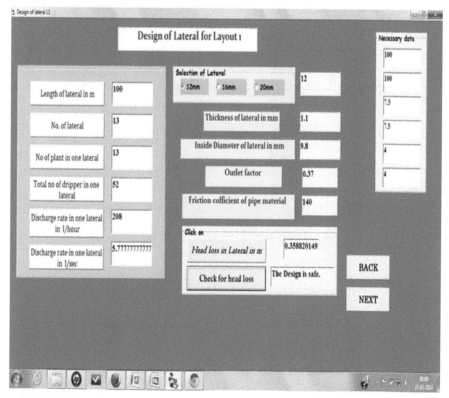

FIGURE 12.11   Design of lateral for layout 1.

The lateral pipe design is followed by design for submain and main line pipes. The water from the water sources is obtained from main line and it is distributed to submain. The main parameter calculated here also includes the head loss at main line pipes. Similar to lateral pipes design, only the diameter of the main line and sub main pipes are to given as input. Other necessary data will be provided by the model like length, discharge of submain and mainline. Then there is a 'NEXT' button to proceed and a 'BACK' button. If user wants to correct any value in the previous page, then 'BACK' button is to be clicked and verified. Figure 12.12 shows the window box for computation of main and sub main pipe for layout 1.

The user is now ready with specification of different pipes to be used in the drip irrigation. The design now focuses on the design of pump system

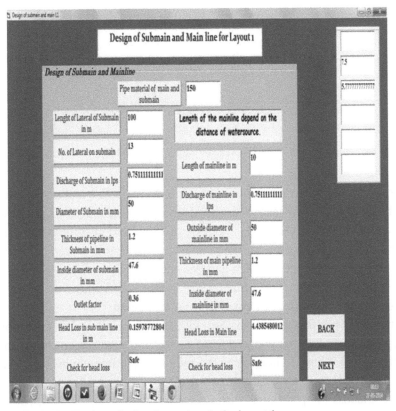

FIGURE 12.12   Design of submains and main for layout 1.

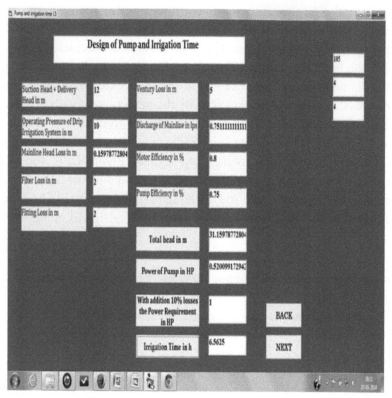

FIGURE 12.13   Design of pump and irrigation time for layout 1.

required to pump necessary amount of water to main line and hence laterals. The design will be based on the head loss at different pipes. The screen shot below shows the necessary input parameters and also the output that are obtained for the power requirement of pump. In this screen irrigation time can also be calculated which decides about the operation of the whole system, to supply water as per plant need. Figure 12.13 shows the window box for design of pump and irrigation time for layout 1.

Once all the design parameters are obtained, the model gives the output of all parameters in a summarized form as shown in the screen shot below (Figure 12.14). This will help the user to view all parameters in one screen instead of navigating through number of screens.

The discussion pertaining design parameter obtained for layout 1 remains same for layout 2 as well. The only difference is the setup of pipes in the field.

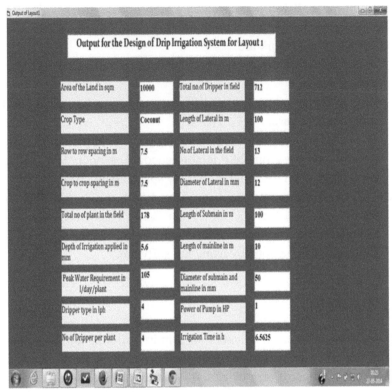

FIGURE 12.14   Output screen of design of drip irrigation system for layout 1.

12.5   CONCLUSIONS

Drip irrigation is becoming popular day-by-day due to efficient use of water. Design of drip irrigation system is the main part of the system. Design of drip irrigation needs lots of mathematical calculations which are complicated and time consuming. It is not so easy to understand by a farmer. To make popularize the drip irrigation system among the farmers, the design part should be simplified. The use of technology is increasing rapidly in agriculture sector. For easy and simple design of drip irrigation system a model need to be develop which can be prove as a helpful tool in agriculture field.

Keeping the above facts, the project was undertaken to study about the design of drip irrigation system. User friendly software was developed for design purpose. It can be proved as a helpful tool, to provide all the possible information regarding components of drip irrigation. The salient findings of the project work are:

- A user friendly software has been developed for design of drip irrigation system.
- Crop water requirement of different crops can be calculated by using the software.
- The software gives the idea about design of drip irrigation system components for crops like lemon, coconut, sapota. The software can also be used for different widely spaced fruit crops.
- Two types of layout have been solved with the help of this software.
- Design for the components like dripper, lateral, submain, main line and pump can be done by using the software.

12.6 SUMMARY

Now-a-days, wide spread application of drip irrigation is highly accepted by the farmers. The success of drip irrigation depends on the proper design of the system. Design of drip irrigation is complicated and time consuming. Hence there is a need to have simplified drip irrigation model to be developed for efficient design of drip irrigation. Design of drip irrigation for fruit and orchard crops can be done easily with use of a model with minimum time for different area and different sizes of orchard.

The present study was undertaken to provide user friendly software for design of drip irrigation system. The objectives of the study were: (i) to compute crop water requirement of different crops using software which can be irrigated by drip irrigation system; and (ii) to develop user friendly software for design of drip irrigation system. This software was developed to design the drip irrigation including computation of water requirement of the crop, design of various units/accessories of the system and power requirement of the system. Three crops ware were taken as study crops. These are sapota, lemon and coconut. Two types of layouts were selected in the study. Though the software uses three crops as mentioned above, it can be modified slightly for other crops to compute the above-mentioned parameters of the drip irrigation system.

KEYWORDS

- coconut
- crop coefficient
- crop spacing
- delivery head
- drip irrigation
- drippers
- emission uniformity of drip system
- emitters
- evapotranspiration
- friction coefficient
- friction loss
- head loss
- horse power
- irrigation requirement
- laterals
- layout
- lemon
- main
- model
- pan coefficient
- pan evaporation
- peak water requirement
- pump
- sapota
- software
- sub main
- suction head
- valve
- visual basic
- water requirement
- wetted area

REFERENCES

1. Arbat, G., Puig-Bargues, J., Duran-Ros, M., Barragan, J., & Cartagena, F. R. (2013). Drip-Irriwater: Computer software to simulate soil wetting patterns under surface drip irrigation. *Computers and Electronics in Agriculture, 98,* 183–192.
2. Behera, A., Kumar, A., & Sethi, R. R. (2013). Simple design software for drip-irrigation system. *Natural Resource Conservation Emerging Issues and Future Challenges,* 161–172.
3. Bombale, V. T., Popale, P. G., & Magar, A. P. (2011). Development of software designof drip irrigation system. *Irrig. Sci., 4*(2), 170–175.
4. Cook, F. J., Thorburn, P. J., Fitch, P., & Bristow, K. L. (2003). Wet Up: Software tool to display approximate wetting patterns from drippers. *Irrig Sci,* 22, 129–134.
5. Feng, J., & Wu, I. P. (1990). *Simple computerized drip irrigation design.* ASAE Publication 04–90, American Society of Agricultural Engineers, St. Joseph, Michigan.
6. Mane, M. S., Ayare, B. L., & Magar, S. S. (2006). *Principle of Drip Irrigation System.* Jain Brothers.
7. Michael, A. M. (2008). *Irrigation Theory and Practices.* Vikas Publication, New Delhi.
8. Molina-Martíneza, J. M., & Ruiz-Canales, A. (2009). Pocket PC software to evaluatedrip irrigation lateral diameters with on-line emitters. *Computers and Electronics in Agriculture, 69,* 112–115.
9. Pedras, C. M. G., & Pereira, L. S. (2001). A simulation model for design and evaluation of micro-irrigation systems. *Irrigation and Drainage, 50*(4), 323–334.
10. Rajput, T. B. S., & Patel, N. (2003). *DRIPD: Software for designing drip irrigation system.* Indian Agricultural Research Institute, New Delhi, TB-ICN:2/2003, 42 pp.
11. Schwab, G. O., Fangmeir, D. D., Elliot, W. J., & Frevert, F. (1992). *Soil Water Conservation Engineering.* John Wiley & Sons, Inc., Fourth Edition.
12. Srivastava, R. C., Verma H. C., Mohanty, S., & Pattanaaik, S. K. (2003). Investment decision model for drip irrigation system. *Irrig Sci,* 22, 79–85.
13. www.en.wikipedia.org/wiki/irrigation_in_india.
14. www.osufacts.okstate.edu.
15. Yurdem, H., Demir, V., & Degirmencioglu, A. (2008). Development of a mathematical model to predict head losses from disc filters in drip irrigation systems using dimensional analysis. *Biosystems Engineering, 100,* 14–23.
16. Zhengying, W., Yiping, T., & Bingheng, L. (2003). A rapid manufacturing method for water-saving emitters for crop irrigation based on rapid prototyping and manufacturing. *Int J Adv Manuf Technol, 21,* 644–648.

APPENDIX I

DRIP IRRIGATION DESIGN CODE BASED ON VISUAL BASIC 6.0.

Form 1
```
Private Sub Command1_Click()
Form2.Show
End Sub
```

Form 2
```
Private Sub cmd_bck1_Click()
Form1.Show
End Sub
Private Sub cmd_cwr_Click()
txtresult3.Text = (Val(txtrr) * Val(txtcc) * Val(txtpe) * Val(txtpc) *
Val(txtkc) * Val(txtwt)) / (Val(txteu))
End Sub
Private Sub cmd_dpth_Click()
txtresult2.Text = Val(txtpe.Text) * Val(txtkc.Text) * Val(txtpc.Text)
End Sub
Private Sub cmd_nxt1_Click()
Form12.txt6.Text = txtresult2.Text
Form12.txt7.Text = txtresult3.Text
Form12.txt1.Text = txt4.Text
Form12.txt2.Text = txt1.Text
Form12.txt3.Text = txtrr.Text
Form12.txt4.Text = txtcc.Text
Form11.txta.Text = txtresult3.Text
Form10.txt1.Text = txtrr.Text
Form10.txtlsm2.Text = txt3.Text
Form10.txt3.Text = txt2.Text
Form9.txt1.Text = txt3.Text
Form9.txt2.Text = txtcc.Text
Form8.txt6.Text = txtresult2.Text
Form8.txt7.Text = txtresult3.Text
```

```
Form8.txt1.Text = txt4.Text
Form8.txt2.Text = txt1.Text
Form8.txt3.Text = txtrr.Text
Form8.txt4.Text = txtcc.Text
Form7.txta.Text = txtresult3.Text
Form6.txtlsm1.Text = txt2.Text
Form6.txt2.Text = txtrr.Text
Form5.txt1.Text = txt2.Text
Form5.txt2.Text = txt3.Text
Form5.txt3.Text = txtrr.Text
Form5.txt4.Text = txtcc.Text
Form3.txt1.Text = txt4.Text
Form3.txt4.Text = txtresult1.Text
Form3.Show
End Sub
Private Sub Command1_Click()
txtresult1.Text = Val(txtrr) * Val(txtcc)
End Sub
Private Sub txt3_Change()
txt4.Text = Val(txt2) * Val(txt3)
End Sub
Private Sub txtcc_Change()
txtresult1.Text = Val(txtrr) * Val(txtcc)
End Sub
Private Sub txtkc_Change()
txtresult2.Text = Val(txtpe) * Val(txtpc) * Val(txtkc)
End Sub
```

Form 3
```
Private Sub cmd_bck2_Click()
Form2.Show
End Sub
Private Sub Command1_Click()
Form4.Show
End Sub
Private Sub cmd_nxt2_Click()
Form12.txt5.Text = txttp.Text
```

```
Form12.txt8.Text = txtlph.Text
Form12.txt9.Text = txtempp.Text
Form12.txt10.Text = txtresult4.Text
Form11.txtb.Text = txtlph.Text
Form11.txtc.Text = txtempp.Text
Form8.txt5.Text = txttp.Text
Form8.txt8.Text = txtlph.Text
Form8.txt9.Text = txtempp.Text
Form8.txt10.Text = txtresult4.Text
Form7.txtb.Text = txtlph.Text
Form7.txtc.Text = txtempp.Text
Form9.txt3.Text = txtempp.Text
Form9.txt4.Text = txtlph.Text
Form5.txt5.Text = txtempp.Text
Form5.txt6.Text = txtlph.Text
Form4.Show
End Sub
Private Sub cmd_td_Click()
txtresult4.Text = Val(txtempp.Text) * Val(txttp.Text)
End Sub
Private Sub cmd_tp_Click()
Y = (Val(txt1.Text) / Val(txt4.Text))
If Y – Round(Y, 0) > 0.5 Then
txttp.Text = Round(Y, 0) + 1
Else
txttp.Text = Round(Y, 0)
End If
End Sub
Private Sub opt1_Click()
txtlph.Text = 2
End Sub
Private Sub opt2_Click()
txtlph.Text = 4
End Sub
Private Sub opt3_Click()
txtlph = 8
End Sub
```

Form 4

```
Private Sub cmd_bck2_Click()
Form3.Show
End Sub
Private Sub cmd_l1_Click()
Form5.Show
End Sub
Private Sub cmd_l2_Click()
Form9.Show
End Sub
Private Sub cmd_nxt2_Click()
Form5.Show
End Sub
```

Form 5

```
Private Sub cmd_chkhl_Click()
If Val(txthl1.Text) < 1.1 Then
txtchkhl1.Text = "The Design is safe."
Else
txtchkhl1.Text = "The Design is not safe."
End If
End Sub
Private Sub cmd_hl1_Click()
txthl1.Text = ((1.21 * (10 ^ 10)) * ((Val(txtqlpsl1.Text) / Val(txtc1.Text))
^ 1.852) * ((Val(txtid1.Text)) ^ (–4.871)) * Val(txtll1.Text) * Val(txtof1.
Text)) / 10
End Sub
Private Sub cmd_ll1_Click()
txtll1.Text = Val(txt2.Text)
End Sub
Private Sub cmd_nl1_Click()
p = (Val(txt1.Text) / Val(txt3.Text))
If p – Round(p, 0) > 0.5 Then
txtnl1.Text = Round(p, 0) + 1
Else
txtnl1.Text = Round(p, 0)
```

```
End If
End Sub
Private Sub cmd_npl1_Click()
X = (Val(txt1.Text) / Val(txt4.Text))
If X – Round(X, 0) > 0.5 Then
txtnpl1.Text = Round(X, 0) + 1
Else
txtnpl1.Text = Round(X, 0)
End If
End Sub
Private Sub cmd_nxt3_Click()
Form8.txt11.Text = txtll1.Text
Form8.txt12.Text = txtnl1.Text
Form8.txt13.Text = txtld1.Text
Form6.txt3.Text = txtqlpsl1.Text
Form6.Show
End Sub
Private Sub cmd_qlph11_Click()
txtqlph11.Text = Val(txttdl1.Text) * Val(txt6.Text)
End Sub
Private Sub cmd_qlps11_Click()
txtqlps11.Text = Val(txtqlph11.Text) / 3600
End Sub

Private Sub cmd_tdl1_Click()
txttdl1.Text = Val(txt5.Text) * Val(txtnpl1.Text)
End Sub
Private Sub Command1_Click()
Form6.txt3.Text = txtqlpsl1.Text
Form6.Show
End Sub
Private Sub Command2_Click()
Form4.Show
End Sub
Private Sub opt1_Click()
txtld1 = 12
```

```
End Sub
Private Sub opt2_Click()
txtld1 = 16
End Sub
Private Sub opt3_Click()
txtld1 = 20
End Sub
Private Sub txtld1_Change()
txtid1.Text = Val(txtld1.Text) – (2 * Val(txttd1.Text))
End Sub
```

Form 6
```
Private Sub cmd_chkhlm1_Click()
If Val(txthm1.Text) < 0.9 Then
txtchkhlm1.Text = "Safe"
Else
txtchkhlm1.Text = "Not safe"
End If
End Sub
Private Sub cmd_chkhls1_Click()
If Val(txthsm1.Text) < 0.9 Then
txtchkhls1.Text = "Safe"
Else
txtchkhls1.Text = "Not safe"
End If
End Sub
Private Sub cmd_hm1_Click()
txthm1.Text = (1.2 * (10) ^ 10) * ((Val(txtqm1.Text) / Val(txtc1.Text)) ^
1.852) * ((Val(txtidm1.Text)) ^ (–4.871)) * Val(txtlm1.Text)
End Sub
Private Sub cmd_hsm1_Click()
txthsm1.Text = (1.2 * (10) ^ 10) * ((Val(txtqsm1.Text) / Val(txtc1.
Text)) ^ 1.852) * ((Val(txtidsm1.Text)) ^ (–4.871)) * Val(txtlsm1.Text) *
Val(txtfsm1.Text)
End Sub
Private Sub cmd_idm1_Click()
txtidm1.Text = Val(txtodm1.Text) – (2 * Val(txttm1.Text))
```

```
End Sub
Private Sub cmd_idsm1_Click()
txtidsm1.Text = Val(txtodsm1.Text) – (2 * Val(txtt1.Text))
End Sub
Private Sub cmd_nlsm1_Click()
p = Val(txtlsm1.Text) / Val(txt2.Text)
If p – Round(p, 0) > 0.5 Then
txtnlsm1.Text = Round(p, 0) + 1
Else
txtnlsm1.Text = Round(p, 0)
End If
End Sub
Private Sub cmd_nxt4_Click()
Form8.txt14.Text = txtlsm1.Text
Form8.txt15.Text = txtlm1.Text
Form8.txt16.Text = txtodsm1.Text
Form7.txt3.Text = txthsm1.Text
Form7.txtq.Text = txtqsm1.Text
Form7.Show
End Sub
Private Sub cmd_qm1_Click()
txtqm1.Text = Val(txtqsm1.Text)
End Sub
Private Sub cmd_qsm1_Click()
txtqsm1.Text = Val(txtnlsm1.Text) * Val(txt3.Text)
End Sub
Private Sub Command13_Click()
Form6.Show
End Sub
Private Sub Command14_Click()
Form7.Show
End Sub
```

Form 7
```
Private Sub cmd_it_Click()
txtresult4.Text = (txta.Text) / (txtb.Text * txtc.Text)
End Sub
```

```
Private Sub cmd_nxt5_Click()
Form8.txt17.Text = txtresult3.Text
Form8.txt18.Text = txtresult4.Text
Form8.Show
End Sub
Private Sub cmd_pump_Click()
txtresult2.Text = (Val(txtq.Text) * Val(txtresult1.Text)) / (75 * Val(txt7.
Text) * Val(txt8.Text))
End Sub
Private Sub cmd_pumphp_Click()
s = Val(txtresult2.Text) * 1.1
If s – Round(s, 0) > 0.5 Then
txtresult3.Text = Round(s, 0) + 1
Else
txtresult3.Text = Round(s, 0)
End If
End Sub
Private Sub cmd_th_Click()
txtresult1.Text = Val(txt1.Text) + Val(txt2.Text) + Val(txt3.Text) +
Val(txt4.Text) + Val(txt5.Text) + Val(txt6.Text)
End Sub
Private Sub Command4_Click()
Form8.Show
End Sub
```

Form 9

```
Private Sub cmd_chkhl2_Click()
If Val(txthl2.Text) < 1.1 Then
txtchkhl2.Text = "Safe."
Else
txtchkhl2.Text = "Not safe."
End If
End Sub
Private Sub cmd_hl2_Click()
txthl2.Text = ((1.21 * (10 ^ 10)) * ((Val(txtqlpsl2.Text) / Val(txtc2.Text))
^ 1.852) * ((Val(txtid2.Text)) ^ (–4.871)) * Val(txtll2.Text) * Val(txtof2.
Text))
```

```
End Sub
Private Sub cmd_ll2_Click()
txtll2.Text = (Val(txt1.Text) / 2)
End Sub
Private Sub cmd_npl2_Click()
p = (Val(txtll2.Text) / Val(txt2.Text))
If p – Round(p, 0) > 0.5 Then
txtnpl2.Text = Round(p, 0) + 1
Else
txtnpl2.Text = Round(p, 0)
End If
End Sub
Private Sub cmd_nxt_Click()
Form12.txt11.Text = txtll2.Text
Form12.txt13.Text = txtld2.Text
Form10.txt2.Text = txtqlpsl2.Text
Form10.Show
End Sub
Private Sub cmd_qlphl2_Click()
txtqlphl2.Text = Val(txttdl2.Text) * Val(txt4.Text)
End Sub
Private Sub cmd_qlpsl2_Click()
txtqlpsl2.Text = Val(txtqlphl2.Text) / 3600
End Sub
Private Sub cmd_tdl2_Click()
txttdl2.Text = Val(txtnpl2.Text) * Val(txt3.Text)
End Sub
Private Sub opt1_Click()
txtld2.Text = 12
End Sub
Private Sub opt2_Click()
txtld2.Text = 16
End Sub
Private Sub opt3_Click()
txtld2.Text = 20
End Sub
```

```
Private Sub txtld2_Change()
txtid2.Text = Val(txtld2.Text) – (2 * Val(txttd2.Text))
End Sub
```

Form 10

```
Private Sub cmd_chkhm2_Click()
If Val(txtchkhm2.Text) < 0.9 Then
txtchkhm2.Text = "Safe"
Else
txtchkhm2.Text = "Not safe"
End If
End Sub
Private Sub cmd_chkhs2_Click()
If Val(txtchkhs2.Text) < 0.9 Then
txtchkhs2.Text = "Safe"
Else
txtchkhs2.Text = "Not safe"
End If
End Sub
Private Sub cmd_hm2_Click()
txthm2.Text = (1.2 * (10) ^ 10) * ((Val(txtqm2.Text) / Val(txtc2.Text)) ^
1.852) * ((Val(txtid2.Text)) ^ (–4.871)) * Val(txtlm2.Text)
End Sub
Private Sub cmd_hs2_Click()
txths2.Text = (1.2 * (10) ^ 10) * ((Val(txtqsm2.Text) / Val(txtc2.Text)) ^
1.852) * ((Val(txtid2.Text)) ^ (–4.871)) * Val(txtlsm2.Text) * Val(txtof2.
Text)
End Sub
Private Sub cmd_lm2_Click()
txtlm2.Text = Val(txt3.Text) / 2
End Sub
Private Sub cmd_nls2_Click()
p = (Val(txtlsm2.Text) / Val(txt1.Text))
If p – Round(p, 0) > 0.5 Then
txtnls2.Text = Round(p, 0) + 1
Else
```

```
txtnls2.Text = Round(p, 0)
End If
End Sub
Private Sub cmd_nxt_Click()
Form12.txt12.Text = txtnls2.Text
Form12.txt14.Text = txtlsm2.Text
Form12.txt15.Text = txtlm2.Text
Form12.txt16.Text = txtod2.Text
Form11.txt3.Text = txthm2.Text
Form11.txtq.Text = txtqm2.Text
Form11.Show
End Sub
Private Sub cmd_qm2_Click()
txtqm2.Text = Val(txtqsm2.Text)
End Sub
Private Sub cmd_qsm2_Click()
txtqsm2.Text = Val(txttnls2.Text) * Val(txt2.Text)
End Sub
Private Sub cmd_tnls2_Click()
txttnls2.Text = 2 * Val(txtnls2.Text)
End Sub
```

Form 11

```
Private Sub cmd_ith_Click()
txtresult4.Text = (txta.Text) / (txtb.Text * txtc.Text)
End Sub
Private Sub cmd_nxt_Click()
Form12.txt17.Text = txtresult3.Text
Form12.txt18.Text = txtresult4.Text
Form12.Show
End Sub
Private Sub cmd_pump_Click()
txtresult2.Text = (Val(txtq.Text) * Val(txtresult1.Text)) / (75 * Val(txt7.
Text) * Val(txt8.Text))
End Sub
Private Sub cmd_pumphp_Click()
```

```
s = Val(txtresult2.Text) * 1.1
If s – Round(s, 0) > 0.5 Then
txtresult3.Text = Round(s, 0) + 1
Else
txtresult3.Text = Round(s, 0)
End If
End Sub
Private Sub cmd_thl_Click()
txtresult1.Text = Val(txt1.Text) + Val(txt2.Text) + Val(txt3.Text) +
Val(txt4.Text) + Val(txt5.Text) + Val(txt6.Text)
End Sub
```

CHAPTER 13

PLANNING, LAYOUT AND DESIGN OF DRIP IRRIGATION SYSTEM

K. CHOUDHARI

CONTENTS

13.1   INTRODUCTION

Drip irrigation is the slow, precise application of water and nutrients directly to the root zones in a predetermined pattern using a point source. It saves water and fertilizer by allowing water to drip slowly to the roots, either onto the soil surface or directly onto the root zone, through a network of valves, pipes, tubing, and emitters. It is done through narrow tubes that deliver water directly to the base of the plant [7].

Presently, the problem facing the world is not the development of water resources, but their management in a sustainable manner. The need of the day is to economize water in agriculture and to bring more area under irrigation, reduce the cost of irrigation on unit land and increase the yield per unit area and unit quantum of water [10, 14]. This can be achieved only by introducing advance irrigation methods like micro irrigation. This when done will not only improve the water productivity, but will also result in arresting the water logging and secondary salinization problems of the canal command areas and check the receding water table and deteriorating water quality in the command areas.

The modern methods of irrigation have number of advantages over the conventional irrigation methods like border, check basin, furrow or surge irrigation. If we could convert sizeable part of irrigated areas into modern irrigation systems, considerably more area can be brought under irrigation along with increasing the land and water productivities [10].

With drip irrigation, water is conveyed under pressure through a pipe system to the fields, where it drips slowly onto the soil through emitters or drippers which are located close to the plants. Compared to other types of irrigation (sprinkler irrigation or surface irrigation), only the immediate root zone of each plant is wetted. Therefore this can be a very efficient method of irrigation. Drip irrigation is sometimes called trickle irrigation [5, 7]. Drip irrigation can be a very technical irrigation system for food or plant production fields. But compared to other technical systems (e.g., sprinkler irrigation) it is a low-technique solution. Drip irrigation requires little water compared to other irrigation methods. The small amount of water reduces weed growth and limits the leaching of plant nutrients down in the soil.

This chapter introduces technology of drip/trickle or micro irrigation.

13.2   HISTORICAL DEVELOPMENT OF DRIP IRRIGATION

Primitive drip irrigation has been used since ancient times. Fan Sheng Chih Shu, written in China during the first century BCE, describes the use of buried, unglazed clay pots filled with water as a means of irrigation [1]. Modern drip irrigation began its development in Germany in 1860 when researchers began experimenting with subsurface irrigation using clay pipe to create combination irrigation and drainage systems [7]. An important breakthrough was made in Germany way back in 1920 when perforated pipe drip irrigation was introduced [1, 3].

The usage of plastic to hold and distribute water in drip irrigation was later developed in Australia [13]. In the United States, the first drip tape, called Dew Hose, was developed by Richard Chapin of Chapin Watermatics in the early 1960s. Modern drip irrigation has arguably become the world's most valued innovation in agriculture since the invention of the impact sprinkler in the 1930s, which offered the first practical alternative to surface irrigation. During the early 1940's Symcha Blass, an engineer from Israel, observed that a big tree near a leaking tap exhibited more vigorous growth than other trees in the area. This led him to the concept of an irrigation system that would apply water in small quantity literally drop by drop. Around 1948, greenhouse operators in the UK began to try a similar method with some modifications. The earliest drip irrigation system consisted of plastic capillary tubes of small diameter (1 mm) attached to large pipes. One of the refinements made by Blass in his original system was coiled emitter. In the early 1960's, experiments in the Israel reported spectacular results when they applied the Blass system in the desert area of the Negev and Arava Drip irrigation pipes began to be sold outside Israel in 1969 on commercial basis. Drip irrigation unit in their current diverse forms were installed widely in USA, Australia, Israel, Mexico and to a lesser extent in Canada, Cyprus, France, Iran, New Zealand, UK, Greece and India [3, 7, 8, 10, 11, 13]

In India drip irrigation was practiced through indigenous methods such as perforated earthenware pipes, perforated bamboo pipes and pitcher/porous cups. In Meghalaya some of the tribal farmers are using bamboo drip irrigation system for betel, pepper and areca nut crops by diverting hill streams in hill slopes [10]. Earthenware pitchers and porous cups have been used for growing vegetable crops in Rajasthan and Haryana. In India drip irrigation was introduced in the early 70's at agricultural universities and other research institutions. The growth of drip irrigation has really gained momentum in the last one decade.

Drip irrigation may also use devices called micro spray heads, which spray water in a small area, instead of dripping emitters. These are generally used on tree and vine crops with wider root zones. Subsurface drip irrigation (SDI) uses permanently or temporarily buried dripper line or drip tape located at or below the plant roots. It is becoming popular for row crop irrigation, especially in areas where water supplies are limited or recycled water is used for irrigation. Careful study of all the relevant factors like land topography, soil, water, crop and agro-climatic conditions are needed to determine the most suitable drip irrigation system and components to be used in a specific installation.

These developments have taken place mainly in areas of acute water scarcity and in commercial/horticultural crops, such as coconut, grapes, banana, fruit trees, and sugarcane; and plantation crops in the states of Maharashtra, Andhra Pradesh, Karnataka, Tamil Nadu and Gujarat.

13.3   CLASSIFICATION OF MICRO IRRIGATION SYSTEMS

13.3.1   DRIP IRRIGATION SYSTEM

In this type drippers/emitters are fitted or pre-fitted at determined spacing in order to cover the root zone of crop and deliver water mostly in the form of drops. It is mostly suitable for widely spaced crops such as Mango, Orange, Grapes, Pomegranate, Coconut, Banana etc.

13.3.1.1   Online Drip Irrigation System

In this system drippers/emitters are pre-fitted on the outer side of the lateral. Spacing between emitter/dripper depends upon the type of the crop. The emitting devices are located in the root zone area. The emitters and laterals are laid on the ground surface along the row. The system is better suited to row crops or widely spaced crops like Mango, Guava, and Pomegranate etc., and horticultural crops.

13.3.1.2   Inline Drip Irrigation System

In this system, drippers/emitters are pre-fitted or inserted or welded at regular intervals into the tubing during the process of production itself and water comes out in the form of drops or ooze and forms a continuous wetting strips in the soil surface or subsurface around the root zone of the crop. The system is better suited to row crops or closely spaced crops like Sugarcane, Tomatoes, Cotton, and Flowers, etc. Such systems are generally preferred in semi-permanent/permanent installations.

13.3.2   MICRO JET/SPRAYER IRRIGATION SYSTEM

In this System water is applied in the form of fine spray in full or part circle on the surface at very low height less 0.5 m or low angle through air around the crops. It does not incorporate any moving parts and has a higher discharge rate and coverage than drippers. The system is suitable for horticultural crops such as Mango, Orange, Lime Coconut, etc., It is also suitable in sandy soil where infiltration rate of water is very high.

13.3.3   MICRO/MINI SPRINKLER SYSTEM

It is just like Micro Jet, and sprays water at height less than 1 m and it incorporates moving parts and thus has greater discharge rate and large coverage range than drippers and micro jets. Hence it is suitable for irrigation of nurseries, lawn, and grass and widely spread canopy crops. Also it is suitable in sandy soil where infiltration rate is very high.

13.3.4   BUBBLER SYSTEM

In this system the water is applied to the soil surface in a small stream or fountain. Bubbler systems do not require elaborate filtration systems.

These are suitable in situations where large amount of water need to be applied in a short period of time and suitable for irrigating trees with wide root zones and high water requirements.

13.3.5 PULSE SYSTEM

Uses high discharge rate emitters and consequently has short water application time. The primary advantage of this system is a possible reduction in the clogging problem.

The comparative irrigation efficiencies under different methods of irrigation is given in Table 13.1, and some of the important differences between modern and other methods of irrigation is given in Table 13.2 [7, 10, 14].

13.4 COMPONENTS OF DRIP IRRIGATION SYSTEM [4]

The pump unit takes water from the source and provides the right pressure for delivery into the pipe system. The control head consists of valves to control the discharge and pressure in the entire system. It may also have filters to clear the water. Common types of filters include screen filters

TABLE 13.1   Irrigation Efficiencies (%) under Different Irrigation Methods

Irrigation efficiency	Methods of Irrigation		
	Surface	Sprinkler	Drip
Application efficiency	60–70	70–80	90
Surface water moisture evaporation	30–40	30–40	20–25
Conveyance efficiency	40–50 (canal)	100	100
	60–70 (well)		
Overall efficiency	30–35	50–60	80–90

Source: Ref. [15].

TABLE 13.2 Comparative Performance of Conventional Irrigation with Micro Irrigation [7, 10, 14]

Performance Indicator	Conventional Irrigation methods	Micro Irrigation
Water saving	Waste lot of water. Losses occur due to percolation, runoff and evaporation	40–70% of water can be saved over conventional irrigation methods. Runoff and deep percolation losses are nil or negligible.
Water use efficiency	30–50%, because losses are very high	80–95%
Saving in labor	Labor engaged per irrigation is higher than drip	Labor required only for operation and periodic maintenance of the system
Weed infestation	Weed infestation is very high	Less wetting of soil, weed infestation is very less or almost nil.
Use of saline water	Concentration of salts increases and adversely affects the plant growth. Saline water cannot be used for irrigation	Frequent irrigation keeps the salt concentration within root zone below harmful level
Diseases and pest problems	High	Relatively less because of less atmospheric humidity
Suitability in different soil Type	Deep percolation is more in light soil and with limited soil depths. Runoff loss is more in heavy soils	Suitable for all soil types as flow rate can be controlled
Water control	Inadequate	Very precise and easy
Efficiency of fertilizer use	Efficiency is low because of heavy losses due to leaching and runoff	Very high due to reduced loss of nutrients through leaching and runoff water

TABLE 13.2   *(Continued)*

Soil erosion	Soil erosion is high because of large stream sizes used for irrigation.	Partial wetting of soil surface and slow application rates eliminate any possibility of soil erosion
Increase in crop yield	Non-uniformity in available moisture reducing the crop yield	Frequent watering eliminates moisture stress and yield can be increased up to 15–150% as compared to conventional methods of irrigation

Source: Refs. [10, 16].

and sand filters which remove materials suspended in the water. Some control head units contain a fertilizer or nutrient tank. These slowly add a measured dose of fertilizer into the water during irrigation. This is one of the major advantages of drip irrigation over other methods. Mainlines, submains and laterals supply water from the control head into the fields. They are usually made from PVC or polyethylene hose and should be buried below ground because they easily degrade when exposed to direct solar radiation. Lateral pipes are usually 12–32 mm in diameter. Emitters or drippers are devices used to control the discharge of water from the lateral to the plants. They are usually spaced more than one meter apart with one or more emitters used for a single plant such as a tree. For row crops more closely spaced, emitters may be used to wet a strip of soil. Many different emitter designs have been produced in recent years. The basis of design is to produce an emitter which will provide a specified constant discharge which does not vary much with pressure changes, and is not easily clogged. A drip irrigation system with components [4] has been shown in Fig. 12.1 in chapter 12.

13.4.1 HEAD CONTROL UNIT

13.4.1.1 Pump/Overhead Tank

It is required to provide sufficient pressure in the system. Centrifugal pumps are generally used for low-pressure trickle systems. Overhead tanks can be used for small areas or orchard crops with comparatively lesser water requirements.

13.4.1.2 By Pass Valves

It is used to bypass the excess flow of water and also reduce the pressure in system. It is installed near the pump of system.

13.4.1.3 Air Release Valves

Air release valves recommended in micro/drip irrigation systems as a safety valve to remove entrapped air and to break the vacuum in the system.

13.4.1.4 Fertilizer Applicator

Application of fertilizer into pressurized irrigation system is done by either a bypass pressure tank, or by venturi injector or direct injection system [7].

13.4.1.5 Filters

The hazard of clogging necessitates the use of filters (Figure 13.1) for efficient and trouble free operation of the micro irrigation system.

a. Centrifugal filters

Centrifugal filters are effective in filtering sand, fine gravel and other high density materials from well or river water. Water is introduced tangentially at the top of a cone and creates a circular motion resulting in a centrifugal force,

Fig. (a) Different types hydro cyclone/centrifugal filters.

Fig. (b) Different types of media filters.

Fig. (c) Screem filter showing steel wire mesh strainers.

Fig. (d) Disk filter showing stacks of discs.

FIGURE 13.1   Types of filters [4].

which throws the heavy suspended particles against the walls. The separated particles are collected in the narrow collecting vessel at the bottom.

b. Gravel or media filters

Media filters (Figure 13.1) consist of fine gravel or coarse quartz sand, of selected sizes (usually 1.5–4 mm in diameter) free of calcium carbonate placed in a cylindrical tank. These filters are effective in removing light suspended materials, such as algae and other organic materials, fine sand and silt particles. This type of filtration is essential for primary filtration of irrigation water from open water reservoirs, canals or reservoirs in which algae may develop. Water is introduced at the top, while a layer of coarse gravel is put near the outlet bottom. Reversing the direction of flow and opening the water drainage valve will clean the filter. Pressure gauges are placed at the inlet and at the outlet ends of the filter to measure the head loss across the filter. If the head loss exceeds more than 30 kPa, filter needs back washing.

c. Screen filters

Screen filters are always installed for final filtration as an additional safeguard against clogging. While majority of impurities are filtered by sand filter, minute sand particles and other small impurities pass through it. The screen filter contains screen strainer, which filters physical impurities and allows only clean water to enter into the micro irrigation system. The screens are usually cylindrical and made of noncorrosive metal or plastic material. These are available in a wide variety of types and flow rate capacities with screen sizes ranging from 20 mesh to 200 mesh. The aperture size of the screen opening should be between one seventh and one tenth of the orifice size of emission devices used.

d. Disk filters

Disk filter contains stacks of grooved, ring shaped disks that capture debris and are very effective in the filtration of organic material and algae. During the filtration mode, the disks are pressed together. There is an angle in the alignment of two adjacent disks, resulting in cavities of varying size and partly turbulent flow. The sizes of the groove determine the filtration grade. Disk filters are available in a wide size range (25,400 microns). Back flushing can clean disk filters. However, they require back flushing pressure of 2 to 3 kg/cm^2.

13.4.1.6   Pressure Relief Valves

These valves may be installed at any point where possibility exists for excessively high pressures, either static or surge pressures. A bye-pass arrangement is simplest and cost effective method to avoid problems of high pressures instead of using costly pressure relief valves.

13.4.1.7   Non Return Valves (NRV)

These valves are used to prevent unwanted flow reversal. They are used to prevent damaging back flow from the system to avoid return flow of chemicals and fertilizers from the system into the water source itself to avoid contamination of water source.

13.4.1.8   Flow Meter

A flow meter measures the flow rate or quantity of water moving through a pipe. In many developed countries, water meters are used to measure the volume of water used by residential and commercial building that are supplied with water by a public water supply system. Water meters can also be used at the water source, well, or throughout a water system to determine flow through a particular portion of the system.

13.4.2   DISTRIBUTION NETWORK

It mainly constitutes of main line, submains line and laterals with drippers and other accessories.

13.4.2.1 Mainline

The mainline transports water within the field and distribute to submains. Mainline is made of rigid PVC and High Density Polyethylene (HDPE). Pipelines of 65 mm diameter and above with a pressure rating 4 to 12 kg/cm^2 are using main pipes.

13.4.2.2 Submains

Submains distribute water evenly to a number of lateral lines. For submain pipes, rigid PVC, HDPE or LDPE (Low Density Polyethylene) of diameter ranging from 32 mm to 75 mm having pressure rating of 2.5 kg/cm^2 are used.

13.4.2.3 Laterals

Laterals distribute the water uniformly along their length by means of drippers or emitters. These are normally manufactured from LDPE and LLDPE. Generally pipes having 12, 16, 20 and 32 mm internal diameter with wall thickness varying from 1 to 3 mm are used as laterals (Figure 13.2).

13.4.2.4 Emitters/Drippers

They function as energy dissipaters, reducing the inlet pressure head (0.5 to 1.5 atmospheres) to zero atmospheres at the outlet. The commonly used drippers are online pressure compensating or online non-pressure compensating, inline dripper, adjustable discharge type drippers, vortex type drippers and micro tubing of 1 to 4 mm diameter (Figure 13.2). These are manufactured from Polypropylene or LLDPE.

a. Online pressure compensating drippers

A pressure compensating type dripper supplies water uniformly on long rows and on uneven slopes. These are manufactured with high quality

Fig. (a) Lateral bundle   Fig. (b) Online pressure compensating drippers

Fig. (c) Online pressure non compensating drippers

FIGURE 13.2   Lateral roll and types of drippers [4].

flexible rubber diaphragm or disc inside the emitter that it changes shape according to operating pressure and delivers uniform discharge. These are most suitable on slopes and difficult topographic terrains.

b. Online non-pressure compensating drippers

In such type of drippers, discharge tends to vary with operating pressure. They have simple thread type, labyrinth type, zigzag path, vortex type flow path or have float type arrangement to dissipate energy. However they are cheap and available in affordable price.

c. *Inline dripper*

These are fixed along with the line, i.e., the pipe is cut and dripper is fixed in between the cut ends, so that it makes a continuous row after fixing the dripper. They have generally a simple thread type or labyrinth type flow path. Such types of drippers are suitable for row crops. Inline tubes are available which include inline tube with cylindrical dripper, inline tubes with patch drippers, or porous tapes or biwall tubes. They are provided with independent pressure compensating water discharge mechanism and extremely wide water passage to prevent clogging. Other accessories are takeout/starter, rubber grommet, end plug, joints, tees, manifolds etc.

13.4.2.5   Valves (ball valve, flush valve)

Ball valve used to sure water does not run out the end of the drip tube and flush valve used to over time a layer of sediment develops inside the tube and needs to be flushed out.

13.4.2.6   End Plug

It is used to close the lateral at ends. It can serve as end cap/manual flush valve by just bending over the end of the drip tubing on itself to crimp off the flow.

13.4.2.7   Pressure Gage

A pressure gauge measures the internal pressure and/or vacuum of a vessel or system. Pressure gages are offered in a variety of styles, sizes and wetted part materials to meet the demands of standard and special applications. In drip irrigation, pressure gage is used to observe the pressure of water at different point in the system.

13.5   ADVANTAGES AND DISADVANTAGES

13.5.1   ADVANTAGES OF DRIP IRRIGATION [7]

- Fertilizer and nutrient loss is minimized due to localized application and reduced leaching.
- Water application efficiency is high if managed correctly.
- Field leveling is not necessary.
- Fields with irregular shapes are easily accommodated.
- Recycled non-potable water can be safely used.
- Moisture within the root zone can be maintained at field capacity.
- Soil type plays less important role in frequency of irrigation.
- Soil erosion is lessened.
- Weed growth is lessened.
- Water distribution is highly uniform, controlled by output of each nozzle.
- Labor cost is less than other irrigation methods.
- Variation in supply can be regulated by regulating the valves and drippers.
- Fertigation can easily be included with minimal waste of fertilizers.
- Foliage remains dry, reducing the risk of disease.
- Usually operated at lower pressure than other types of pressurized irrigation, reducing energy costs.
- Minimum diseases and pest infestation.
- Usually operated at lower pressure than other types of pressurized irrigation, reducing energy costs.

13.5.2   DISADVANTAGES OF DRIP IRRIGATION [7]

- Initial cost can be more than overhead systems.
- The sun can affect the tubes used for drip irrigation, shortening their usable life.
- If the water is not properly filtered and the equipment not properly maintained, it can result in clogging.
- Drip tape causes extra cleanup costs after harvest. Users need to plan for drip tape winding, disposal, recycling or reuse.

- Waste of water, time and harvest, if not installed properly. These systems require careful study of all the relevant factors like land topography, soil, water, crop and agroclimatic conditions, and suitability of drip irrigation system and its components.
- Most drip systems are designed for high efficiency, meaning little or no leaching fraction. Without sufficient leaching, salts applied with the irrigation water may build up in the root zone, usually at the edge of the wetting pattern. On the other hand, drip irrigation avoids the high capillary potential of traditional surface applied irrigation, which can draw salt deposits up from deposits below.
- The PVC pipes often suffer from rodent damage, requiring replacement of the entire tube and increasing expenses.

13.6 APPLICATIONS OF DRIP IRRIGATION

Drip irrigation is being used in farms, commercial greenhouses, and residential gardeners (Figure 13.3). Drip irrigation is adopted extensively in areas of acute water scarcity; and especially for crops and trees such as coconuts, containerized landscape trees, grapes, bananas, ber, eggplant, citrus, strawberries, sugarcane, cotton, maize, and tomatoes.

Drip irrigation kits for garden are increasingly popular for the homeowner and consist of a timer, hose and emitter. Hoses that are 4 mm in diameter are used to irrigate flower pots.

13.7 PLANNING AND SURVEY FOR DESIGN OF DRIP IRRIGATION SYSTEM

The planning, survey and design of drip irrigation system is an essential step to supply the required amount of irrigation water. The water requirement of the plant per day depends on the water that is taken by the plant from the soil and the amount of water evaporating from the soil in the immediate vicinity of the root zone in a day. The plant intake is affected by the leaf area, stage of growth, climate, soil conditions, etc., The water requirement and irrigation scheduling can be determined from the soil or plant indicators based methods or soil water budget method, but the sim-

Fig. (a) Drip irrigation in grapes

Fig. (b) Drip irrigation in banana

Fig. (c) Nursery flowers watered with drip irrigation

Fig. (d) Pot Irrigation by on-line drippers

Fig. (e) Drippers for soilless growing trays

Fig. (f) Inline drip irrigation system;
Image on right from <http://www.efreshglobal.com>

FIGURE 13.3   Examples of drip irrigation.

plest and most common method is to use USDA class A pan data. To supply the required amount of water uniformly to all the plants in the field, it is essential to design the system to maintain desired hydraulic pressure in the pipe network. The design of micro irrigation system is essentially a decision regarding selection of emitters, laterals and manifolds, sub main, main pipeline and required pumping unit [7].

13.7.1   SURVEY

For correct depiction of shape and slope of any plot of land, a typical lay-out (Figure 13.4) is considered and the following elements are essentially required:

   a. Straight distance between end points as AB, BC, CD, DA (Figure 13.4).

   b. Angles as necessary at A or B or C or D are required.

   c. Elevations when the ground is undulating or slope is more than 1%.

13.7.2   PROCEDURE

- Before starting the work, it is necessary that a reconnaissance of the area is done by the survey engineer and the boundary of the plot it fixed by putting stones on all the end points. This will ensure correct alignment for distance measurement and measurement for the lengths for angles.
- Distance:
  - I. Distance may be measured with a tape, preferably a 30 meter tape.
  - II. Distance should be measured in a straight line with signals duly erected at the point to which distance is to be measured.

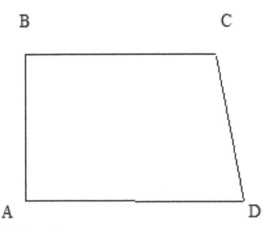

FIGURE 13.4   Field layout.

- Angles:
  I.   Angles as necessary should be measured by measuring tie length between the lines forming the angles. First measure 10 m distance along the two sides and mark arrows/pegs. Now measure the length of Tie between the arrows/pegs.
  II. For a three cornered figure no angle need to be measured.
  III. For a four cornered closed figure, only angle needs the measures.
  IV. The angle can be determined using the following equation.
       i. $\theta = 2\sin^{-1}(x/20)$, where θ = angle to be measured, x = tie length
       ii. The angle can also be determined using Table 13.3.
       iii. Check the traverse using following equation:
       Sum of all internal angles = $(2n - 4) \times 90$, where n is the number of sides.
- Elevations:
  I.   The slope of the ground may be judged with eye where possible, and a remark regarding this should be entered in the plan.
  II. Where the ground is flat or the difference of heights between end points is less than ½%, a remark regarding this should be entered in the plant.
  III. Where the ground is undulating and the height differences cannot be judged with eye, levels should be taken with the help of a leveling instrument. Assuming the height of a permanent point like well to be 100.00 meters, heights of other points should be deducted and entered in the plan.

13.7.3   WATER SOURCE, PUMP, EXISTING PIPE LINE AND OUTLETS

- The position of the water source, (Well, Bore, Reservoir, River) existing pipe line and the outlet from which water is to be used for the drip system, should be accurately marked on the survey plan. Following details regarding the outlet should be entered:
  I.   Size of outlet and its height above ground level.

TABLE 13.3   Angles for Tie Length

Tie (Meters)	Angle (Degrees)	Tie (Meters)	Angle (Degrees)
10.0	60.00	13.6	85.6
10.1	60.66	13.7	86.47
10.2	61.32	13.9	88.05
10.4	62.66	14.0	88.85
10.5	63.33	14.1	89.65
10.6	64.1	14.15	90.00
10.7	64.68	14.2	90.46
10.8	65.36	14.3	91.28
10.9	66.04	14.4	92.10
11.0	66.72	14.5	92.93
11.1	67.42	14.6	93.77
11.2	68.11	14.7	94.61
11.3	68.80	14.8	95.46
11.4	69.50	14.9	6.31
11.5	70.19	15.0	97.18
11.6	70.90	15.1	98.05
11.7	71.60	15.2	98.92
11.8	72.31	15.3	99.81
11.9	73.02	15.4	100.70
12.0	73.73	15.5	101.61
12.1	74.45	15.6	102.52
12.2	75.17	15.7	103.44
12.3	75.90	15.8	104.37
12.4	76.63	15.9	10.31
12.5	77.36	16.0	106.26
12.6	78.10	16.2	108.19
12.8	79.58	16.3	109.17
12.9	80.33	16.4	110.16
13.0	81.08	16.5	110.1
13.1	81.83	16.6	112.19
13.2	82.59	16.7	113.23
13.3	83.36	16.8	114.28
13.4	84.13	16.9	115.34
13.5	84.90	17.0	116.42

II. Details and description of the outlet, whether threaded, flange type of only pipe.

III. Whether of GI, PVC or Alkathene pipe. These details will enable the designer to list down the fittings for filter etc.

- In addition, the following details regarding the pump should be ascertained and entered in the questionnaire form:

  I.   Suction head and size of suction pipe.

  II. Delivery head and size of delivery pipe.

  III. Time for which the pump runs.

  IV. For bore wells, the following extra data need be collected.

- Diameter of bore
- Bore depth
- Stages of submersible pump
- Delivery head and size of delivery pipe
- A note regarding the farmer's choice for location of filter station, where necessary, should be entered on the questionnaire form.

13.7.4. PERMANENT DETAILS

Any permanent details like a huge rock, a farm house, a large tree etc., falling inside/close to the plot, should be surveyed. Take angular measurements from at least two end points so that these may be plotted by their angles and shown in the survey plan.

13.7.5 WATER AND SOIL SAMPLES

Water and soil samples should be collected for analysis in our laboratory and for any special advice regarding suitability of soil/water for some particular crops. Water samples (at least 750 ml) should be collected after the pump has been running for at least 10 minutes. Soil sample (About 1 kg) should be collected from about four places well spread in the field and from a depth of about 1 foot below ground level.

13.7.6 SURVEY PLAN

From the above data, a survey plan should be prepared on 1:1000 scale and all details should be shown in the plan.

13.7.7   SPECIAL POINTS

Any other special points to which the farmer wants to draw attention, e.g.:
*   Name for billing, date by which quotation is required.
*   Any advice which the farmer requires regarding suitability of land for some particular crop or choice of a pump should be specifically noted and entered in the questionnaire.

13.7.8   CONCLUSIONS

*   Survey work requires extra care as any wrong observation/recording will make the design work difficult/erroneous.
*   All entries in the questionnaire form should be therefore be made very carefully and accurately.
*   Only an accurate survey can result in accurate design.

13.8   DESIGN OF MICRO IRRIGATION SYSTEMS

Appropriate design of micro irrigation system is very essential to obtain proper performance and benefits. Each irrigation system should be designed taking into consideration all agro climatic factors, crop physiology, soil characteristics, water source and other engineering factors. Objectives of design are:
*   To maintain higher system and irrigation efficiency by means of higher emission uniformity.
*   To maintain optimum moisture level in soil for optimization of crop yield.
*   To keep both initial investment and annual cost at minimum level.
*   To design a suitable type of system, which will last and perform well.
*   To design a manageable system, which can be easily operated and maintained.
*   To satisfy and fulfill the requirements of crops and farmers or users.

13.8.1   DESIGN INPUTS

As we are aiming at a precise quantity and uniform application of water for each and every plant, collection of data as detailed below is a prerequisite for designing an efficient micro irrigation system:

- Engineering Survey: Measurement of field, ground slope, contours.
- Water Source: Assessment of water source and availability of water.
- Agricultural Details: Crop, spacing, type, variety, age, water requirement, crop physiology.
- Climatological data: Temperature, humidity, rainfall, evaporation, etc.
- Soil & Water Analysis: Collection of soil & water samples and analyzing.

Singh et al. [12] have indicated the details of steps that can be followed by the designer under following headings:

- a. System capacity.
- b. Selection of emitting devices or drippers or tubings.
- c. Selection and design of laterals or tubes.
- d. Selection and design of submains.
- e. Selection and design of mainlines.
- f. Selection and design of filtration system.
- g. Selection and design of pump unit.

Tables 13.4 summarizes peak flow rate for different crops to achieve better irrigation performance and better crop yield.

13.8.2   HEAD LOSS IN LATERALS

The pipes used in micro-irrigation system are made of plastics and considered as smooth pipe. The pressure drop due to friction can be evaluated with the help of Hazen -William empirical equation as below:

$$H_f (100) = K \times [Q/C]^{1.852} \times [D]^{-4.871} \times F \qquad (1)$$

$$F = 1/[m+1] + 1/[2N] + [m-1]^{0.5}/[6 N^2] \qquad (2)$$

TABLE 13.4 Peak Water Requirements for Different Crops [5–7, 10]

Crop	Spacing (ft × ft)	Peak water requirement (Lit/Day/Plant)
Ber	10×10	38
	12×12	55
Coconut	25×25	80
Cotton/Groundnut	In-line at 6 ft	12 lit/m/day
Custard apple	10×10	38
	12×12	54
Grapes	6×4	12
	8×6	18
	8× 8	24
	8×10	30
Guava	15×15	80
	18×18	100
	25×25	130
Orange/Lemon/ Citrus	16×16	75
	18×18	85
Papaya	5×4	16
	7×7	18
Pomegranate	10×10	30
	12×12	40
	15×15	70
Sapota/Mango	25×25	120
	30×30	170
Sugarcane	In line at 8 ft	18 lit/m/day
	In line at 7 ft	16 lit/m/day
	In line at 6 ft	14 lit/m/day
Vegetables	In line at 6 ft	13 lit/m/day

where, H (100) = head loss due to friction per 100 meter of pipe length, m/100 m; Q = flow of water in pipe, lps; D = internal diameter of pipe, cm; L = length of the pipe, m; C = Hazen–William constant (140 for PVC pipe); F = reduction factor due to multiple openings in pipe, which can be computed by Eq. (2). As the length of the pipe increases, the discharge in the pipe decreases due to emission outlets and hence the total energy drop is less than as given by the above equation. For this reason, a reduction factor F is introduced; m = 1.852; and N = number of outlets on the lateral.

The design criteria for lateral pipe are to keep pressure variation and discharge variation within the prescribed limit. For lateral design, the discharge and operating pressure at the emitter are required to be known. Based on this, the allowable head loss can be calculated using above formula. The diameter of lateral pipe is usually selected so that the difference in discharge between emitter operating simultaneously will not exceed 10%. Pressure head difference should not exceed 10 to 15% of the operating pressure. For the discharge variation of 10%, the emission uniformity has to be more than 90%.

The submains line hydraulics is similar to that of the lateral hydraulics. The submain hydraulic characteristics can be computed by assuming the laterals are analogous to emitters on lateral line. Hydraulic characteristics of submain and mainline pipe are usually taken hydraulically smooth since PVC and HDPE pipe are normally used. The Hazen-William roughness coefficient is usually taken between 140 and 150. The energy loss in the submain can be computed with the methods similar to those used for lateral computations.

The size of mainline is determined by considering the quantity of water flowing through it, length and path or mainline, elevation of ground, velocity, safety parameters, cost economy and nomograms provided by the manufacturer. Usually the pressure controls or adjustments are provided at the submain inlet. Therefore, energy losses in the main line should not affect system uniformity. There is no outlet in case of mainline therefore reduction factor is not multiplied. The frictional head loss in main pipeline is calculated by the Hazen-William formula.

The selection and design of filtration system is based on:
- Source of water
- Type, size and concentration of physical impurities.
- Designed System flow (filtration capacity)
- Type of Irrigation System.
- Workability of filtration System.
- Ease for handling, cleaning, maintenance and repairing.

- Filtration media and low frictional losses.
- Economical investment, maintenance and power cost.

Pump unit is a electromechanical device, which lifts water from one level to another level with pressure. Total head required for the system is calculated as [12]:

Total head = (Suction + Delivery) head + Filtration losses
                        + Frictional losses in main line + Operating pressure
                        + Fitting losses + Venturi head + elevation head (if any) (3)

With total head and discharge required, one can calculate power required by the pump for efficient operating of micro irrigation system [12].

$$HP = [Q \times H]/[75 \times a \times b] \qquad (4)$$

where, Q = Discharge required in liters per second; H = Total head required in meter; a = Efficiency of motor (assumed 85%); b = Efficiency of pump. (assumed as 80%); and HP = Calculated horse power. One should refer the manufacturer chart to know the actual horse power of pump.

13.9   WATER REQUIREMENT OF CROPS

Before calculating crop water requirements, following points are to be taken into consideration.

- Type of crop and its age.
- Type of soil and wetting pattern (Figure 13.5).
- Evaporation loss from the surface.
- Transpiration loss from leaves.
- Canopy area and root zone development.
- Plant-to-plant and row-to-row spacing.
- Wind velocity, humidity, etc.

After studying all above factors, the month-wise and age-wise water requirement of the crop is decided and accordingly system is designed. As a first step in the proper design of the irrigation system, it is necessary to know the crop water requirements. In general terms, the crop water requirement is equivalent to the rate of evapotranspiration necessary to

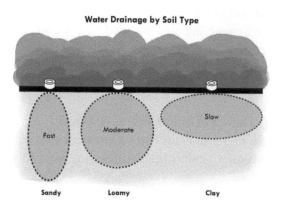

FIGURE 13.5 Wetting pattern below the dripper: left – Heavy soil, Center – Medium soil, and Right – Light soil.

sustain optimum plant growth. The accuracy of the determination of crop water requirement will be largely dependent on the type of climatic data available. In order to calculate the water requirement of crops accurately one should know the following variables:

13.9.1 CROP STAGE/AGE OF PLANT

Water required by a plant varies with its growth stage. Water requirement is different at the time of sowing, when the plant is growing, at flowering stage, at the time of fruiting, ripening, and harvesting.

13.9.2 SATURATION CAPACITY OR MAXIMUM WATER HOLDING CAPACITY

Under these conditions, the soil is fully saturated and all soil pores are filled with water. In this state, plant roots get suffocated due to absence of air in the root zone and cannot uptake water properly.

13.9.3 FIELD CAPACITY

It is defined as the amount of water held in soil after excess water has drained away and the moisture content has become relatively stable. At field capacity, large soil pores are filled with air and the micro pores are filled with water. The soil moisture tension at this stage generally varies from 0.1 to 0.3 atmospheres and roots can uptake water and nutrients with ease. In drip irrigation, the soil moisture content is always maintained near field capacity level.

13.9.4   PERMANENT WILTING POINT

It is the soil moisture content at which plants can no longer obtain enough water to meet transpiration needs and remain wilted even if water is added to the soil.

13.9.5   TRANSPIRATION

It is the evaporation of water from plant surfaces directly into the atmosphere or into intercellular spaces and then by diffusion through the stomata to the atmosphere.

13.9.6   EVAPOTRANSPIRATION (ET)

It is also called consumptive use. It denotes the quantity of water transpired by plants during their growth or retained in the plant tissues plus the moisture evaporated from the surface of the soil and the vegetation.

13.9.7   REFERENCE CROP EVAPOTRANSPIRATION (ET0)

It is the rate of evapotranspiration from an extended surface of 8 to 15 cm tall green grass cover of uniform height, actively growing, completely shading the ground and not short of water.

13.9.8   CROP FACTOR OR CROP COEFFICIENT (KC)

The crop factor, Kc, is selected for the given crop and stage of crop development under prevailing climatic condition. For each crop, there are four growth stages. Crop factors have to be determined for each of the stage. The crop factor for the initial stage is lower (0.3 to 0.4) and it increases to (0.7 to 0.8) during the crop development stage. It is about 1.0 to 1.10 at fully grown stage and it reduces again to 0.8 to 0.9 at the harvesting stage.

13.9.9   CANOPY FACTOR

The canopy factor indicates the growth of crop at different stages. It is expressed as the ratio of the area covered by plant foliage to the total area provided for the plant. Thus, it is the ratio of the plant's shadow area at noon to the area of the plant. The area provided to a plant is the product of plant spacing and row spacing.

13.9.10   ESTIMATION OF CROP WATER REQUIREMENT

13.9.10.1   Tree Crops

$$\text{Daily water required, liters per tree} = [A \times B \times C \times D]/[E] \qquad (5)$$

In Eq. (5): A = evapotranspiration rate in mm/day; B = crop factor; C = canopy factor = area of plant shadow at 12 noon/ (plant spacing x row spacing); D = area of plant = (plant spacing x row spacing); and E = efficiency of irrigation system (0.9 for drip system and 0.8 for sprinkler system). The daily water requirement for each crop is calculated considering the age of the crop and the temperature and evapotranspiration data from Meteorological department for that region. After knowing the daily water requirement, the time of operating the drip system is decided depending on the flow capacity of online or inline drip system.

Numerical example 1: Given: Crop = Grapes, Row Spacing = 1.8 m, Row Spacing = 1.2 m, Age = 5 to 6 years, Place: Pune District, Month: May; A = Evapotranspiration rate (mm/day) = 8 mm/day in summer in Pune district, B = Crop factor = 0.7, C = canopy factor = 1.0, D = area of plant = (plant spacing x row spacing) = 1.8 m × 1.2 m, and E = efficiency of irrigation system = 0.9. Calculate crop water requirement under these conditions. Therefore,

Daily water required, liters per tree = $[8 \times 0.7 \times 1 \times 1.8 \times 1.2]/[0.9]$ = 13.44 or 14 liters/day/plant.

The water requirement of grape plant grown in a different location in district of Maharashtra state, for same conditions, will be different because the temperature and evapotranspiration rate in summer will be different at these places than at Pune.

13.9.10.2 Row Crops

Water requirement per square meter for row crops such as Sugarcane, cotton, vegetables, etc., is calculated using Eq. (5), where: D = spacing between two Inline laterals in meters,

Numerical example 2: Crop – Sugarcane, Inline to inline spacing = 1.8 m, Place – Solapur District, Month: May; A = 10, B = 0.7, C = 1, D = 1.8 m, and E = 0.9.

Sugarcane water requirement = $[10 \times 0.7 \times 1 \times 1.8]/[0.9]$ = 14 liters/ meter/day

This implies that the water requirement for 1 m length of sugarcane row is 14 liters every day.

13.10 DESIGN EXAMPLE

Given Data
Crop: Grape Spacing: 8′ × 6′ (feet) = 2.43 × 1.82 m
Area: 100 m × 150 m
Pump Delivery Size: 4"
Electricity Available: 8 hours.
Design a suitable drip irrigation system.

Solution
Step 1: Peak water requirement = PWR = $[A \times B \times C \times D]/[E]$ = $[7 \times 0.7 \times 0.85 \times 8 \times 6]/[0.9 \times 3.28 \times 3.28]$ = 20.64 liters/day/plant

Step 2: Water application rate = WAR = No. of drippers x dripper flow rate = $2 \times 8 = 16$ lph/day

Step 3: Irrigation time = [PWR]/[WAR] = 20.64/16 = 1.29 hours

Step 4: Design and selection of lateral

SDR of lateral = [No. of drippers x dripper flow rate]/[plant-to-plant spacing] = $[2 \times 8] [6/3.28] = 8.74$ lph per m

From SDR Curve, we get the maximum permissible length of lateral: Lateral 12 mm $\varphi = 26$ m; Lateral 16 mm $\varphi = 52$ m.

We select, lateral of 16 mm diameter of running length 50 m.

Step 5: Design and selection of submain

Plant population per submain = Area/spacing = $[100 \times 150]/[(6/3.28) \times (8/3.28)] = 3362$

SDR of submain = [plant population × no. of drippers × dripper flow rate]/[length of submain] = $[3362 \times 2 \times 8]/[100 \times 2] = 269$ lph per m

From SDR Curve, we get the maximum permissible length of lateral of 50 mm $\varphi = 52$ m; and for 63 mm $\varphi = 84$ m

Here, we select 63 mm φ of PVC, Class II (4 Kg/cm^2) submain line.

Frictional Head Losses = 1.4 m in 84 m length.

Step 6: Design and selection of main line

Flow in submain = [plant population x no. of drippers x dripper flow rate]/3600 = $[3362 \times 2 \times 8]/[3600 \times 4] = 3.73$ lps

From rate of flow curves, for Q = 3.73 lps, We select, 75 mm φ, Class II (4 Kg/cm^2) main line.

Friction Loss Calculation:

From	To	Length (m)	Q (lps)	Diameter (mm)	$H_f/1000$ (m)	Actual H_f (m)
W.S	V1	162.5	3.73	75	16	2.6

Total Head Loss (H) = $10 + 5 + 2 + 10 + 2.6 = 34.6$ m

Step 7: Selection and design of pump

HP = $[3.73 \times 34.6]/[75 \times 0.8 \times 0.85] = 2.53 = 3$ HP

Step 8: Selection of filter capacity

Q_{filter} = Flow × [3600/1000] = $3.73 \times 3.6 = 13.42$ m^3/h or take it as 20 m^3/h

Step 9: Irrigation scheduling

Shift	Crop	Time (hour)	Flow (lps)
I	Grape	1.29	3.73
II	Grape	1.29	3.73
III	Grape	1.29	3.73
IV	Grape	1.29	3.73

Step 10: Detailed BOQ for Grape Plantation, in this chapter: 1.00 US$ = 60.00 Rs.

S. No	Item Code	Description of Item	Quantity	Unit	Rate	Amount (Rs.)
1		PVC Pipe 75 mm φ, II (4 Kg/cm²) as per	168	m		
2		PVC Pipe 63 mm φ, II (4 Kg/cm²) as per	210	m		
3		Plain lateral 16 mm φ	6200	m		
4		Grommet Take Off 16 mm φ	168+10	Nos		
5		End Stop '8' Shape 16 mm φ	178	Nos		
6		Dripper (8 lph)	6800	Nos		
7		Ball Valves 63 mm φ	4	Nos		
8		Flush Valves 63 mm φ	4	Nos		
9		Sand Filter 25 m³/hr (2") with plastic manual backwash manifold	1	No		
10		Screen Filter 25 m³/hr (2")	1	No		
11		Air Release Valve (1")	1	No		
12		Air Release Valve Assembly (1" × 63 mm)	1	No		
13		By Pass Assembly (3" x 2.5")	1	No		
14		PVC Elbow 63 mm φ	24	Nos		
15		PVC Tee 75 mm φ	5	Nos		
16		GI Nipples (2")	6	Nos		
17		GI Pipes (2.06)	6	Nos		
18		GI Socket (2")	4	Nos		
19		GI Elbow (2")	3	Nos		
20		GI Pipe (2.012)	4	Nos		
21		PVC Elbow 75 mm φ	1	No		
22		PVC Reducer 75 x 63	1	No		

Step 11: Considerations for cost estimate

- Area unit used should be hectare throughout cost estimate document: Area conversion factor of 2.47 acre = 1 hectare should be used.
- MIS component should be with specification/dimension.
- Use the pipe charts or nomograms for piping design work.
- Item quantity should be estimated based on the site visit/survey/ design prepared.
- Incorporate future requirements of the farmer, if any, in the design.
- All the data and drawings should be mentioned clearly on the design drawing so that they are useful for future use by the farmer and Company's visiting engineers/officers/technicians.
- Area considered for cost estimate should be equal to the design area & in no case should exceed revenue area.
- Indicate North Direction.
- Scale of drawing should be in MKS unit and not in FPS unit.
- In orchard crops, mention loop per plant in meter, no. of drippers per plant and their discharge. In intercropping system, cropping pattern should be shown with row-to-row and plant-to-plant spacing.
- In intercropping, consider average lateral spacing for lateral/ emitting calculation and also mention cropping pattern in design drawing.

13.11 APPLICATION OF FERTILIZERS AND OTHER AGRO CHEMICALS (FERTIGATION AND CHEMIGATION)

Fertigation is the method of application of soluble fertilizer with irrigation water. Fertigation is a prerequisite for drip irrigation. Since the wetted soil volume is limited, the root system is confined and concentrated. The nutrients from the root zone are depleted quickly and a continuous application of nutrients along with the irrigation water is necessary for adequate plant growth. Fertigation offers precise control on fertilizer application and can be adjusted to the rate of plant nutrient uptake [7].

13.11.1 ADVANTAGES OF FERTIGATION

- The supply of nutrients can be more carefully regulated and monitored.
- The nutrients can be distributed more evenly throughout the entire root zone or soil profile.
- The nutrients can be supplied incrementally throughout the season to meet the actual nutritional requirements of the crop.
- Nutrients can be applied to the soil when crop or soil conditions would otherwise prohibit entry into the field with conventional equipment.
- Soil compaction is avoided, as heavy equipment never enters the field.
- Crop damage by root pruning, breakage of leaves, or bending over is avoided, as it occurs with conventional chemical field application techniques.
- Less equipment may be required to apply the chemicals and fertilizers.
- Less energy is required in applying the chemical. Usually less labor is needed to supervise the application.

All chemicals applied through irrigation systems must meet the following criteria:

- Avoid corrosion, softening of plastic pipe and tubing, or clogging of any component of the system.
- Safe for field use.
- Soluble or emulsifiable in water.
- Should not react adversely to salts or other chemicals in the irrigation water.

13.11.2 EQUIPMENT AND METHODS FOR FERTILIZER INJECTION

Injection of fertilizer and other agrochemicals such as herbicides and pesticides into the drip irrigation system is done by Bypass pressure tank, Venturi system or Direct injection system.

a. **Bypass pressure tank** employs a tank into which the dry or liquid fertilizers kept. The tank is connected to the main irrigation line by means of a Bypass so that some of the irrigation water flows through the tank and dilutes the fertilizer solution. This bypass flow is brought about by a pressure gradient between the entrance and exit of the tank, created by a permanent constriction in the line or by a control valve.

b. **Venturi injector**: A constriction in the main water flow pipe increases the water flow velocity thereby causing a pressure differential (vacuum) which is sufficient to suck fertilizer solution from an open reservoir into the water stream. The rate of injection can be regulated by means of valves. This is a simple and relatively inexpensive method of fertilizer application.

c. **Direct injection system**: With this method a pump is used to inject fertilizer solution into the irrigation line. The type of pump used is dependent on the power source. The pump may be driven by an internal combustion engine, an electric motor or hydraulic pressure. The electric pump can be automatically controlled and is thus the most convenient to use. However its use is limited by the availability of electrical power. The use of a hydraulic pump, driven by the water pressure of the irrigation system, avoids this limitation. The injection rate of fertilizer solution is proportional to the flow of water in the system. A high degree of control over the injection rate is possible, no serious head loss occurs and operating cost is low. Another advantage of using hydraulic pump for fertigation is that if the flow of water stops in the irrigation system, fertilizer injection also automatically stops. This is the most perfect equipment for accurate fertigation. Two injection points should be provided, one before and one after the filter for fertigation. This arrangement helps in bypassing the filter if filtering is not required and thus avoids corrosion damage to the valves, filters and filter screens or to the sand media of sand filters. The capacity of the injection system depends on the concentration, rate and frequency of application of fertilizer solution.

13.12   STEPS IN INSTALLATION OF MICRO IRRIGATION SYSTEM

13.12.1   INSTALLATION OF PUMPING UNIT

The centrifugal pump is installed as close to the water surface as possible. It is located at an easily accessible place in clean, dry and well-ventilated surroundings. To ensure maximum capacity, the site selected should permit the use of the shortest and most direct suction and discharge pipes. The foundation should be rigid enough to absorb all vibrations. The pump and driver must be carefully aligned. The suction piping should be as direct and short as possible. It should have minimum of fittings so as to avoid excessive friction losses. The use of bends, elbows, tees and other fittings is kept to the minimum to reduce head loss in the discharge line.

13.12.2   INSTALLATION OF FERTIGATION UNIT

Water soluble fertilizers/chemicals are injected into micro irrigation system through fertilizer tanks, venturi type meter or injection pumps. The fertilizer tank/ venturi injector or injection pump is connected parallel to the irrigation pipe line by creating differential pressure. Non-return valve is installed to prevent contamination of water source.

13.12.3   CONSIDERATIONS FOR INSTALLATION OF FILTRATION UNIT

- Minimum use of fitting such as elbows and bends to be made.
- The filter unit should be fixed on the delivery side of the pump.
- Care should be taken to see that the filter size should be in accordance with the capacity of the system.

13.12.4   INSTALLATION OF MAINS AND SUBMAINS

Except for fully portable system, both mains and sub mains are installed underground at a minimum depth of about 0.5 m such that they are unaffected by cultivation or by heavy harvesting machinery. Even for systems, which have portable laterals that are removed at the end of each season, it is common practice to install permanent underground sub mains. Generally sub mains run across the direction of the rows.

13.12.5   LAYING OF LATERALS

Generally laterals are laid on the ground surface. Usually laterals are placed along contours on sloping field. Burying laterals underground might be necessary or at least have some advantages for some installations. Where this is done, the emission devices should be above ground level. The downstream end of the lateral can be closed by simply folding back the pipe and closing it with a ring of larger diameter pipe, known as end plug. This can be easily slipped for flushing. The simplest connection for low-pressure system is for the lateral to be inserted directly into the sub main. Slightly undersized hole in the sub main is cut with the help of twist meter drill bit. The hole is expanded with the tapered tool, and then the lateral is inserted quickly after withdrawing the taper. The lateral is cut at an angle of about 450 at the end.

13.12.6   PUNCHING OF LATERALS AND FIXING OF DRIPPERS

- Water is passed through the laterals and flushed so that it gets bulged and makes easy for punching.
- The holes on the lateral are made as per the required spacing.
- The dripper position should be fixed according to design, soil water report and water requirement in peak summer.
- Punching should be done from the sub main.
- While fixing the dripper, the dripper should be pushed inside the lateral and pulled slightly to ensure leak proof connection.
- The end of lateral should be closed with end cap.

13.13   FIELD EVALUATION OF THE MICRO IRRIGATION SYSTEM

- Backwash the filter till clean water comes out through its flush valve.

- See that all the gate valves and flush valves are opened before testing.
- Close the flush valve after the sub main is completely flushed.
- When the laterals are completely flushed, close with the help of end caps.
- Check the pressure on the gauges installed at inlet and outlet of the filter.
- Obtain the desired pressure at the filter, if excess pressure is observed open the bypass valve slowly till the desired pressure is obtained.
- Check the working of air release valve at the submains.
- At this pressure, measure the discharge at a minimum of three places (first, middle and last dripper of lateral) by volumetric method.
- The emission uniformity of microirrigation system can be estimated by using following formula.
- Modify the design/ change drippers, if the Emission Uniformity is less than 85 %.

13.14 MAINTENANCE AND REMEDIES FOR TROUBLESHOOTING IN MICRO IRRIGATION SYSTEM

13.14.1. GENERAL MAINTENANCE

Filter is the heart of a drip system and its failure will lead to clogging of the entire system. Pressure differential across the filter is the correct indication of the timing of cleaning of the filter.

13.14.1.1 Sand Filter

- Backwash the filter daily for five minutes to remove the silt and other dirt accumulated during the previous day's irrigation.
- Do not allow pressure difference across the sand filter more than 0.3 kg/cm^2.

- Once in a week, while backwashing, allow the backwash water to pass through the lid instead of the backwash valves.
- Stir the sand in the filter bed up to the filter candles without damaging them. Whatever dirt is accumulated deep inside the sand bed will get free and goes out with the water through the lid.

13.14.1.2   Screen Filter

- Clean screen filter everyday
- Open the flushing valve on the filter lid so that the dirt and silt will be flushed out.
- Open the filter and take out the filter element and clean them from both sides. Care should be taken while replacing the rubber seals, otherwise they may get damaged.
- Do not allow pressure difference across the screen filter more than 0.2 kg/cm^2.
- Never use hard brush to rub screen surface.

13.14.2   SUBMAIN AND LATERAL FLUSHING

Sometimes silt escapes through the filters and settles in sub mains and laterals. Also some algae and bacteria lead to the formation of slimes/pastes in the pipe and laterals. To remove them, the sub mains should be flushed by opening the flush valves. The lateral lines are flushed by removing the end caps. By flushing, even the traces of accumulated salts will also be removed. The flushing is stopped once the water going out is cleaned.

13.14.3   CHEMICAL TREATMENT

Clogging or plugging of emitters/orifices of bi-wall will be due to precipitation and accumulation of certain dissolved salts like carbonates, bicarbonates, iron, calcium and manganese salts. The clogging is also due to the presence of microorganisms and the related iron and sulfur slimes due to algae and bacteria. The clogging or plugging is usually avoided/cleared by chemical treatment of water. Chemical treatment commonly used in micro irrigation systems includes addition of chloride and/or acid to the water supply. The frequency of chemical treatment depends on degree of problem at the site. As a general rule, acid treatment should be performed once in ten days and chlorine treatment once in fifteen days.

13.14.3.1   Acid Treatment

Hydrochloric acid is injected into drip system at the rates suggested in the water analysis report. The acid treatment is performed till a pH of 4 is observed at the end of lateral length. After achieving a pH of 4 the system is shut off for 24 h. Next day the system is flushed by opening the flush valve and lateral end caps.

13.14.3.2   Chlorine Treatment

Chlorine treatment in the form of bleaching powder is performed to inhibit the growth of microorganism like algae and bacteria. The bleaching powder is dissolved in water and this solution is injected into the system for about 30 minutes. Then the system is shut off for 24 hours. After that the lateral end caps and flush valves are opened to flush out the water with impurities. The bleaching powder can directly be injected through venturi at the rate of 2 mg/l.

13.14.4   DRIP IRRIGATION SYSTEM TROUBLESHOOTING [7]

Problems	Causes	Remedies
Accumulation of sand and debris in screen filter	Displacement of filter element. Less quantity of sand in filters	Place filter element properly. Fill required quantity of sand
Drop in pressure	Leakage in main opened outlet. Low water level in well.	Arrest the leakage and close outlet. Lower the pump with reference to well water level
Leakage of water from air release valve.	Damaged air release valve ring	Replace the damaged ring.

Problems	Causes	Remedies
More pressure at the entry of sand filter	No bypass in the pipeline/ bypass not opened. Displacement of filter element. Less quantity of sand in filters	Provide bypass before filter and regulate pressure. Place filter element properly. Fill required quantity of sand
More pressure drop in filters	Accumulation of dirt in filters	Clean filters every week. Back wash the filters for every 5 minutes daily.
Oily gum material comes out on opening the lateral end	More algae or ferrous material in water	Clean the laterals with water or give chemical treatment
Oily gum material comes out on opening the lateral end	More algae or ferrous material in water	Clean the laterals with water or give chemical treatment
Out coming of white mixture on removing the end plug	More salinity in water. Un cleaned lateral	Remove the end stop. Clean the laterals fortnightly
Pressure gauge not working	Rain water entry inside. Corrosion in gauge pointer damage	Provide plastic cover and fix pointer properly.
Under flow or over flow from laterals	Clogging of drippers. Unclosed end plug	Clean the sand and screen filters. Close the end cap
Venturi not working during chemical treatment and fertigation	Excess pressure on filters Improper fitting of venturi assembly	Bypass extra water to reduce pressure Repair the venturi assembly.
Water not flowing up to lateral end	Holes in laterals. Cuts in laterals. Bents in laterals.	Close the holes and cuts. Remove the bends.

13.15   SUMMARY

Presently in world the problem facing is not just the development of water resources, but their management in a sustainable manner. The need of the day is to economize water in agriculture and to bring more area under irrigation, reduce the cost of irrigation on unit land and increase the yield per unit area and unit quantum of water. This can be achieved only by introducing advance irrigation methods like micro irrigation.

Micro irrigation unit in their current diverse forms have been installed widely in U.S.A., Australia, Israel, Mexico and to a lesser extent in Canada, Cyprus, France, Iran, New Zealand, UK, Greece and India. In

India drip irrigation is being practiced through indigenous methods such as perforated earthenware pipes, perforated bamboo pipes and pitcher/porous cups. After that micro irrigation was introduced in the early 70's at agricultural universities and other research institutions. The growth of micro irrigation has really gained momentum in the last one decade. These developments have taken place mainly in areas of acute water scarcity and in commercial/horticultural crops, such as coconut, grapes, banana, fruit trees, sugarcane and plantation crops.

The modern methods of irrigation have number of advantages over the conventional irrigation methods. If we could convert sizeable part of irrigated areas into modern irrigation systems, considerably more area can be brought under irrigation along with increasing the land and water productivities.

Drip irrigation is the slow, precise application of water and nutrients directly to the plant's root zones in a predetermined pattern using a point source. It saves water and fertilizer by allowing water to drip slowly to the roots of many different plants, either onto the soil surface or directly onto the root zone, through a network of valves, pipes, tubing, and emitters. It is done through narrow tubes that deliver water directly to the base of the plant.

This book chapter is specially focused on the design and of drip irrigation system because appropriate design of drip irrigation system is very essential to obtain proper performance and benefits. Each Irrigation System should be designed taking into consideration of agro climatic factors, crop physiology, soil characteristics, water source and other Engineering factors. Careful studies of all the relevant factors are needed to determine the most suitable drip irrigation system and components to be used in a specific installation. Maintenance of drip irrigation system plays a key role to get 100% performance of system which while gives higher output over traditional irrigation system.

KEYWORDS

- **advantages**
- **agricultural details**
- **chemigation**
- **classification**

- climatological data
- components
- crop water requirement
- design of drip irrigation system
- disadvantages
- drip irrigation
- drip irrigation survey
- drip irrigation system
- dripper
- fertigation
- fertilizer injection method
- filters
- head loss
- historical development
- inline dripper
- installation
- irrigation efficiency
- laterals
- mainline
- maintenance
- micro irrigation system
- online dripper
- soil and water analysis
- submains
- testing
- troubleshooting
- water resources management

REFERENCES

1. David A. Bainbridge, (2001). Buried clay pot irrigation: a little known but very efficient traditional method of irrigation. *Agricultural Water Management, 48*(2), 79–88.

2. Council WSSCC (Water Supply and Sanitation Collaborative Council) (2016). Drip Irrigation [Accessed: 15 February]. http://www.sswm.info/category/implementation-tools/water-use/hardware/optimisation-water-use-agriculture/drip-irrigation

3. Drip Irrigation — History and Benefits. www.ideorg.org.

4. Drip Irrigation catalog, Jain Irrigating System Limited.

5. FAO (1988). Irrigation Water Management: Irrigation Methods. Food and Agriculture Organization of the United Nations, Rome, http://www.fao.org/.

6. Food and Agriculture Organization of the United Nations (2010). Crop water requirements. FAO, Rome: URL [Accessed: 7 December]. http://www.fao.org/.

7. Goyal, Megh R. (2012). *Management of Drip/Trickle or Micro Irrigation*. Oakville, Canada: Apple Academic Press, Inc., ISBN 9781926895123, p. 104.

8. History of drip irrigation: http://www.gardenguides.com/79735historydripirrigation.html.

9. INFONET-BIOVISION (2010). *Water for Irrigation*. Zürich: Biovision. URL [Accessed: 02.08]. http://www.infonet-biovision.org.

10. Narayanamoorthy, A. (2008). Potential for drip and sprinkler irrigation in India. http://nrlp.iwmi.org/PDocs/DReports/Phase_01/12.%20Water%20Savings%20Technologies%20-%20Narayanmoorthy.pdf.

11. Standish, S. (2009). How to irrigate on a shoestring. *Portland, USA: Global Envision*, p. 14.

12. Singh, R. K., Agrawal, K. N., & Satapathy, K. K. (2004). Prospects of drip irrigation in North Hill region. Paper presentation #9, In: Bimal J. Deb, & B. Datta-Ray (eds.), *Proceedings of Changing Agricultural Scenario in North-East India (Shillong)*, Concepts Pub. Co., New Delhi – 110059, pp. 136–149.

13. The History of the Drip Irrigation System and What is Available Now? http://www.irrigation.learnabout.info/articles/138dripirrigationsystem.htm.

14. Tilley, E., Luethi, C., MoreL, A., Zurbruegg, C., & Schertenleib, R. (2008). Compendium of sanitation systems and technologies. *Duebendorf, Switzerland: Swiss Federal Institute of Aquatic Science and Technology (EAWAG) and Water Supply and Sanitation Collaborative, http://www.sandec.ch; http://www.sswm.info; http://www.wsscc.org*, and *http://www.eawag.ch*.

15. Sivanappan, R. K. (1998). Status, scope and future prospects of micro irrigation in India. Proc. Workshop on Micro irrigation and Sprinkler Irrigation System. CBIP New Delhi, April 28–30, p. 17.

16. Sivanappan, R. K. (1994). Status, prospects of micro irrigation in India. *Irrigation and Drainage Systems*, 8, 49–58.

CHAPTER 14

HEAD LOSS IN DOUBLE INLET LATERAL OF A DRIP IRRIGATION SYSTEM

K. S. MOHANTY, B. PANIGRAHI, and J. C. PAUL

CONTENTS

14.1   INTRODUCTION

Globally, fresh water to a tune of 3240 M km³ is being utilized. Of this, 69% is being used in agriculture sector, 8% in domestic and 23% in industrial

This is an edited version: Kirti S. Mohanty, "Evaluation of head loss in a double inlet lateral of a drip irrigation system" (2011). Unpublished thesis for Master of Technology, Department of Soil and Water Conservation Engineering, College of Agricultural Engineering and Technology, Orissa University of Agricultural & Technology, Bhubaneswar. In this chapter: One US$ = 60.00 Rs. (Indian Rupees).

and other sector. In India, around 88% water is being used in agriculture sector, covering around 85 M ha areas under irrigation. Due to liberalization of industrial policies and other developmental activities, the demand for water in industrial and domestic sectors is increasing day-by-day. This forces to reduce the percentage area under irrigation. The growing demand from the population calls for more efforts to enhance agricultural production. The horticulture sector has emerged as a promising area for diversification in agriculture on account of high-income generation for unit area, water and other farm inputs and environmental friendly production systems. Government of India has accorded high priority for development of this sector since VIII plan by enhancing the plan grant of 240 million Rs. in VII plan to 10 billion Rs. [23]. Horticulture crops show promising results when irrigated by micro irrigation system.

14.1.1 MICRO IRRIGATION SYSTEM

Micro irrigation involves frequent application of water directly on or below the soil surface near the root zone of plants. It delivers required and measured quantity of water in relatively small amounts slowly to the individual or groups of plants. Water is applied as continuous drops, tiny streams or fine spray through emitters placed along a low-pressure delivery system. This system provides water precisely to plant root zones and maintains ideal moisture conditions for optimum plant growth. The available literature and the results obtained at Precision Farming Development Centre and Indian Institute of Technology (IIT), Kharagpur and other research centers report that there is 50–70% saving in irrigation water, 18 to 152% increase in yield of fruits and vegetables crops under drip irrigation [23].

14.1.2 SINGLE INLET DRIP IRRIGATION SYSTEM

Single inlet drip irrigation system is the most popular and traditional method of drip irrigation layout. This system consists of a single submain to which laterals are fitted and run along the rows of the crop. Now-a-days, this system is very much adopted in all types of fruits and vegetables.

14.1.3   DOUBLE INLET DRIP IRRIGATION SYSTEM

Research studies on application of irrigation water through double inlet drip irrigation are scarce. Advantages of drip irrigation by double inlet drip system are that it reduces the length of the laterals thereby reducing the frictional head loss in laterals. Since the length of laterals pipes in drip irrigation is more as compared as the length of main and submains, there is considerable savings in head loss in double inlet drip irrigation system. Saving of head loss has an extra advantage that it reduces the pump capacity, thereby reduces the cost of cultivation which ultimately enhances the economics of the system [3].

Another added advantage of double inlet drip system is that it ensures uniform water application in the root zone of crops, which enhances the biometrics and yield of crops and thereby increasing the water application efficiency (WUE). Some theoretical study on single and double inlet drip irrigation system has been done by Nayak [14]. In double inlet system, there is considerable savings of head loss. However, the study has not been conducted in experimental fields to actually verify how much savings of head loss is achieved in double inlet drip system compared to single inlet system. Therefore, it is felt necessary to conduct practical studies on farmer's field. It is also necessary to verify what is the impact of the system on crop yield and on the technological feasibility of the system?

Tomato is a highly remunerative vegetable crop. It is liked by almost all people. Earlier studies report that the crop responses very well to drip irrigation.

Keeping all these facts in view, authors conducted the research study in this chapter with the specific objectives, namely: to calculate of head loss and dripper discharge in double inlet drip irrigation system and compare it with single inlet system; and to study the crop yield in both single inlet and double inlet drip system.

14.2   REVIEW OF LITERATURE

Irrigation is an age-old practice, which is a basic requirement for sustainable agricultural production. There has been continuous development of different irrigation practices, but a revolutionary development of drip irrigation system came which is the result of a large number of studies conducted by scientists, engineers and manufacturers. It has been proven that it is the most efficient technology in the field of pressurized irrigation. In drip irrigation, plants are frequently watered as per the consumptive use of plants, thereby minimizing conventional losses such as deep percolation, runoff and soil water evaporation and thus soil moisture advance remain within desired range. In the following paragraphs the available literatures regarding the above study including head loss in pipe systems especially in laterals have been reviewed.

A study has been made of the effect of the online emitter in the energy losses in trickle irrigation lateral by Ahmed [1]. The study involved eight types of emitters with various barb areas installed in five different commonly used polythene pipes of various diameters. Results indicated that there are significance energy losses due to the emitter connection. The values of these losses are the function of the area of the emitter barb protrusion and the lateral pipe diameter. An increase in the energy loss of more than 32% was found by polythene pipe compared with plain pipe observed for laterals as compared with plain pipe laterals observed. Author suggested a simple procedure to incorporate the emitter barb losses in the design of trickle irrigation. Similar reports are also available in literature [17, 19–21].

A study was undertaken by Bagarello et al. [2] to find out simple pressure parameters such as maximum pressure, minimum pressure and average pressure along the lateral line in a rectangular submain unit which can be used for the hydraulic design of micro irrigation systems. This is based on the fact that simple pressure ratios (minimum pressure to maximum pressure and minimum pressure to average pressure) are all indications of the uniformity of micro irrigation systems. A definite relationship between the total friction loss and maximum and minimum pressure difference, or average and minimum can be determined for a micro irrigation system under different field slope situations. When a nominal pressure head (10 m) is set for the average or maximum pressure, the minimum pressure can be determined based on the selected design criteria. The total friction loss can be considered as the sum of the total friction loss for the lateral and submain. The length of the lateral and the size of submain can be determined from the respective total friction pressure losses.

Improved irrigation water use efficiency is an important component for sustainable agricultural production. Efficient water delivery systems such as subsurface drip irrigation (SDI) can contribute immensely towards improving crop water use efficiency and conserving water. However, critical management considerations such as choice of SDI tube, emitter spacing and installation depth are necessary to attain improved irrigation efficiencies and production benefits. In the study, Giuseppe et al. [5] evaluated the effects of subsurface drip tape emitter spacing (15, 20 and 30 cm) on yield and quality of sweet onions grown at two locations in South Texas— Weslaco and Los Ebanos. Season-long cumulative crop evapotranspiration (ET_c) was 513 mm in Weslaco and 407 mm at Los Ebanos. Total crop water input at Weslaco was roughly equal to ET_c whereas at Los Ebanos, water inputs exceeded ET_c by about 35%. Crop water use efficiency was slightly higher at Weslaco (13.7 kg/m^3) than at Los Ebanos (11.7 kg/m^3) because of differences in total water inputs resulting from differences in irrigation management [5].

Lateral lines of a drip irrigation system consist of pressurized pipelines with inline or online emitters. Proper hydraulic design of drip laterals usually requires the accurate evaluation of the total head losses, represented by friction losses along the pipe and the emitters, and local losses due to the emitter connections [6, 7]. The local loss evaluation procedure previously obtained for coextruded laterals on the basis of new experiments were extended. In addition, a simplified procedure was proposed based on the constant outlet discharge assumption for a quick evaluation of total head losses in drip irrigation lines taking into account the total local loss due to the emitter connections. Total head loss values measured on 15 commercially available coextruded laterals was then compared with those obtained by using the newly proposed methodology. Relative errors on the pressure head estimation for the examined cases were found to be always ±2.4%, and therefore the proposed methodology could serve for a quick, approximate evaluation of the total head losses along the laterals [6, 9].

Importance of considering local losses, when a high number of emitters, are installed along the laterals was highlighted by John [7]. For online emitters, local loss is due to the turbulence consequent to the protrusion of emitter barbs into the flow, whereas for inline emitters, whose diameter is usually smaller than the pipe's diameter, local losses are due to both the contraction and the expansion of the flow stream lines at the emitter connections. In the latter case, an additional continuous friction loss due to the diameter being smaller than the pipe's must be considered.

Minor head losses at emitter insertions along drip laterals were studied by Juana et al. [8]. The minor head losses were predicted by a derivation of Be'langer's theorem and analyzed by the classic formula that includes a friction coefficient K multiplied by a kinetic energy term. A relationship was established for K as a function of some emitter geometric characteristics. They take into account the flow expansion behind the reduction of the cross-sectional area of the pipe due to obstruction by the emitter. Flow constrictions at emitter insertions were estimated by analogy with contraction produced by water jets discharging through orifices. An experimental procedure was also developed to determine minor losses in situ in the laboratory or in the field. An approach was suggested to calculate either K or the emitter equivalent length as a function of lateral head losses, inlet head, and flow rate. Internal diameter and length of lateral, emitter spacing, emitter discharge equation, and water viscosity must be known for this purpose. Approximate analytical relations to study flow in laterals were developed. They may be used to design and evaluate drip irrigation units. Analytical and experimental procedures were validated in the companion paper by Juana et al. [8].

A study on design of bi-wall irrigation method needs information on its hydraulics such as pressure-discharge relationship, uniformity coefficient (measure of uniformity of discharge) and head loss along the bi-wall lateral having orifice spacing of 30 cm × 150 cm was selected. The pressure discharge relationship was established for the bi-wall, for operating heads ranging from 3 m to 15 m of water [4].

An interesting theoretical analysis for the water flow from the line sources was studied by John [7]. Basing on the solutions of linearized forms of different flow equations, subjected to certain boundary and initial conditions, they have made their analysis. The solutions produced theoretical matric flux potential and streamline distribution below a line source. Head losses, along lateral lines of drip irrigation systems strongly affect the available head at emitter nozzles. Consequently, discharge distribution was reported to be significantly affected when conventional non compensating emitters were used. These losses are frequently estimated by adding frictional losses, along uniform pipe sections between consecutive emitters to singular minor losses, resulting in form resistance at emitter insertions. He determined the energy drops in the lateral lines of two types of portable drip irrigation units. They are one irrigation unit having one lateral end connected with a container and other end was plugged, i.e., called single inlet drip irrigation and another system having both end of lateral connected with two container, i.e., called double inlet system. He calculate the energy drop (h_{f1}) by friction in the lateral line of the portable single inlet drip irrigation and in the lateral line of the portable double inlet system (h_{f2}) by Hazen–William formula for plastic pipe and derived a relation, $h_{f2} = h_{f1} \times 0.14$ [14].

Provenzano et al. [16] gave the accurate design procedure of drip irrigation laterals, which needs to consider the variation of hydraulic head due to pipe elevation changes, head losses along the lines, and also at a given operating pressure, emitter discharge variations related to manufacturing variability, clogging, and water temperature. Hydraulic head variations were consequent to both the friction losses and local losses due to the in-line or on-line emitters along the pipe, which determine the contraction and subsequent enlargement of the flow streamlines. They reported that, in-line emitters usually have a smaller diameter than the pipe, and therefore an additional friction loss must be considered. Evaluation of energy losses and consequently the design of drip irrigation lines is usually carried out by assuming the hypothesis that local losses can be neglected, even if previous experimental researches showed that local losses can become a significant percentage of total head losses as a consequence of the high number of emitters installed along the lines. This paper reported the results of an experimental investigation to evaluate local losses in integrated laterals in which coextruded emitters were installed inside the pipe. Local losses were measured for 10 different types of commercially available integrated laterals and for different Reynolds numbers. A practical power relationship was deduced between the coefficient, expressing the amount of local losses as a fraction of the kinetic head and a simple geometric parameter characterizing the geometry of the emitter and the pipe. Local losses obtained for integrated laterals were then compared with those due to the on-line emitters, previously determined as a function of the pipe-emitter geometry. The proposed criterion for calculating the local losses was finally verified by using a step-by-step procedure.

A simple method for drip irrigation systems was developed using energy gradient line approach for the tapered pipes in non-uniform slopes [7]. A non-uniform slope tapered pipe flow problem was divided in to a series of uniform slopes and uniform diameter pipe flow sub problems. An equation for determining the dripper discharge rates and determining the discharge of pipe for a uniform slope and uniform diameter pipe has been developed. The dripper discharges for each submain are considered as independent sample and the variation of discharges for the total pipe length was found out using statistical equations.

Experimental investigations were undertaken to evaluate trickle irrigation emitter barb losses for different types of on-line trickle irrigation emitters using 12 mm lateral pipe [18]. Total eight types of on-line trickle emitters of three familiar brands with rated discharge ranging from 2 to 8 lph were studied under different flow velocities ranging from 1.5 to 2.0 m/s at an interval of 0.1 m/s. Darcy's friction factor for 12 mm trickle lateral line was estimated using Blasius equation and was found to be 0.0259. Emitter barb loss in terms of equivalent length of lateral pipe was estimated for all the eight types of emitters under study. Linear relationship between flow velocity and barb loss were developed for each emitter. Relationship also has been developed between the barb protrusion area and barb loss.

An experiment to study the effect of planting geometry on yield, capital cost, operating cost and net return for the banana crop planted in a one-ha area and irrigated using a trickle irrigation system was carried on by [23]. The cost analysis was carried out based on yield results obtained under different planting geometry pattern at 2 m, 3 m, 4 m and 5 m row spacing for one to four plants transplanted separately at a place. The net return was found to be maximum for one plant at a place at 2-m spacing. It was found that the length to breadth ratio of planting had high correlation with the initial capital cost and the total annual cost. The highest return in investments was obtained at 4-m spacing with 2 plants per location. Analysis was carried out to study the effect of market price of banana, water price and irrigation level on benefit cost ratio and net return.

Analytical solution of the optimum hydraulic design problem for micro irrigation submain units of specified dimensions was carried on by [25].

New algebraic equations were derived to calculate explicitly the optimum values of the design variables. The design variables were the lengths of two given pipe sizes for the laterals as well as the appropriate lengths of the available pipe sizes for the manifold. Tapered laterals and manifold were selected in such a way that the sum of the costs of the laterals and the manifold was minimized, while the hydraulic design criterion was ensured. The case of a single-diameter lateral with tapered manifold pipeline was also examined. The design procedure also could be applied in sprinkle irrigation tapered laterals. The explicit optimum design solution was demonstrated in two cases studied.

A new approach for solving lateral hydraulic problems in laminar or turbulent flow was developed by Tiwari [22]. The outflow was treated as a discrete variable event by means of Taylor polynomials used to calculate the flow rates along the lateral (minimal outflow included). The friction head losses were calculated using the Darcy-Weisbach equation with a non-constant logarithmic friction factor, f. This algorithm allowed hydraulic computation for a set of connected laterals (with different pipeline diameter, slope, flow regime, or emitter spacing). The new approach could be used to calculate flow variation on laterals and submains in trickle or sprinkler irrigation systems without an excessive calculation effort. Results were comparable to those obtained in the literature.

A simplified method for the resolution of lateral hydraulic problems in laminar and turbulent flow was proposed by Von Berunath [24, 25]. In the first stage, the head losses were calculated by applying the Darcy–Weisbach equation with a discrete and constant outflow model, which lead to a correction parameter equivalent to Anwar's Ga factor. The difficulty that arises from variation of the friction factor along the lateral due to discharge flow was overcome by means of an equivalent friction factor. In the second stage, this head loss model was used together with a variable discharge model based on Taylor polynomials to make a better estimate of the flow rate distribution by means of a successive-approximations scheme. This new approach directly allowed the computation of the real mean lateral's outflow and the minimum and maximum discharges. In the third stage, the previous results could be improved if desired by taking into account the non-constant outflow distribution model developed in the previous stage. The method proposed was useful to work out the hydraulic computation of laterals with the inlet segment at full or fractional outlet spacing, and complex laterals when a different pipeline diameter, slope, flow regime, or emitter gap was to be considered. The results are comparable to those obtained in the literature.

Wu [26] reported that direct calculations could be made for all emitter flows along a lateral line and in a submain unit based on an energy gradient line (EGL) approach. Errors caused by the EGL approach were evaluated by a computer simulation. A revised energy gradient Line (REGL) approach developed using a mean discharge approximation, could reduce the errors and match with the results from a step-by-step calculation for all emitters in a drip system. The developed equations could be used for computerized design of drip irrigation systems. He reported that the hydraulic design of micro irrigation systems to achieve high system uniformity had led design engineers to over-design irrigation systems arbitrarily. Commonly used emitter flow variations of 10–20% were equivalent to a uniformity coefficient of about 95–98%, or a coefficient of variation of emitter flow of only 3–7%. The uniformity of a micro irrigation system was affected not only by hydraulic design but also manufacturer's variation, grouping of emitters, plugging, soil hydraulic characteristics and emitter spacing. Among all the factors affecting the uniformity, the hydraulic design, with an emitter flow variation of 10–20% produced only a few percent changes in uniformity. The manufacturer's variation of micro irrigation emitters ranged from 2 to 20%. The hydraulic variation would be less significant when an emitter with 10% or more manufacturers' variation was selected. The grouping effect reduced the coefficient of variation to half or more if four or more emitters could be grouped together. The effect of hydraulic design was also less significant with plugging situations. When there was no plugging, the emitter flow variation from 10 to 20% in hydraulic design will reduce spatial uniformity only about 8% from 93% to 85% when the emitter spacing is designed as half of the wetting diameter in the field.

A number of researchers have done extensive researches and contributed significantly in the field of friction loss in drip irrigations. The research studies by famous investigators [10–13, 15, 17] are very important in this field.

14.3   MATERIALS AND METHODS

14.3.1   BASIC DATA FOR THE STUDY AREA

The study was conducted in the village of Kantabada of district Khurda having 20'15" N latitude, 85'50" E longitude and elevation of 46 m above mean sea level. Agriculture, horticulture, and animal husbandry are the predominant economic activities in the village. Although there are two reservoirs, i.e., the Deras and the Jhumuka dam closely situated to this village, yet poor and erratic rainfall coupled with poor water management practice has resulted in a steady depletion of groundwater resources. The hybrids, high-response varieties that react to conditions of plentiful water and chemical nutrition, have failed to make an impact in stressful dry land agricultural conditions. Excessive irrigation for paddy, has led to water stress in the region. The population of the village is around 3000 with around 280 families. Agriculture provides only seasonal employment and the economic returns from the land are low. Non-agricultural economic activities are poorly developed. There is a very good communication system to the village as it is almost equidistance from Cuttack, Bhubaneswar and Khurda Town, i.e., 15 km from Khandagiri chawk towards Sum hospital. This area is only 20 km from the capital city Bhubaneswar of the state of Orissa which is the main town. Most of the agricultural land is composed of rainfed areas with very unpredictable yield. The entire production takes place in the canal-irrigated land in the command areas of the two reservoirs.

The purpose of this study is to recommend new operational approaches in the transition from the traditional paddy cultivation to hi-tech vegetable production with optimal use of water.

14.3.2   CLIMATE

The present experimental study was undertaken in the village Kandabada which is situated in central part of the state of Orissa, India. The study site bears a very pleasant weather suitable for all crops available in the state. In summer this area shows a minimum temperature of 16°C. The minimum temperature in winter never falls below 7°C, with a tropical climate. In monsoon season, the study area receives 1240 mm of rainfall. It experiences typical tropical weather conditions, and succumbs to the heat and cold waves that sweep in from north India. The summer months from March to May are hot and humid and maximum temperature often reaches 45°C in May. The southwest monsoon lashes Orissa in June, bringing relief to the parched environs of this area. July and August receive the maximum rainfall. Monsoon withdraws from the area by first week of October.

14.3.3 SOIL AND TOPOGRAPHY

The study area belongs to laterite soil zone having soil depth 15 cm with slightly undulating land. The pH of soil is 5.4 which is slightly acidic. The soil is typically suitable for vegetable production. The infiltration capacity of soil is about 0.15 cm/h. The mechanical composition of soil in coarse sand, fine sand, silt and clay is 50.3, 22.32, 8.32, and 18.88%, respectively. The soil of the study area belongs to sandy loam in textural class. Bulk density of the soil is 1.53 g/cm³. The fertility status of the soil in very good having Nitrogen, Phosphorus and Potash as 230.00, 13.62, 180.00 kg/ha. Figure 14.1 represents the view of upper soil characteristic of the study area.

14.3.4 TOMATO CROP

The study area suffers from scarcity of water for irrigation purpose during the non-rainy season. The local farmers usually prefer to grow

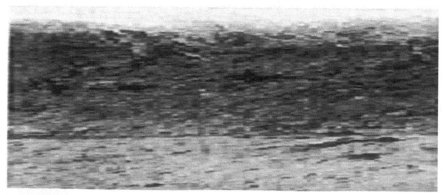

FIGURE 14.1   Soil of the study area.

different vegetables during the *rabi* season by using certain alternate irrigation arrangements viz. drip, sprinkler irrigation systems, etc., The farmers in the area prefer to grow tomato crop since it needs less water and is a highly remunerative crop. Keeping in view, tomato is selected as the study crop in the particular study area. A popularly known variety of tomato namely BT-12 having duration of 120 days was planted with plant-to-plant spacing of 60 cm and row-to-row spacing being maintained also at 60 cm. The variety BT-12 is disease resistant especially with resistant and gives good yield. This variety of tomato does not rot easily due to high temperature or during transportation. It has a hard outer membrane and is suitable for salad.

14.3.5 RAISING TOMATO SEEDLINGS

Seedlings were grown before one month of transplanting in raised beds of 2 m width and of 3 m length. Soil solorization of nursery bed by covering them with white transparent polythene sheet for one month was done in hot summer months, to kill the disease causing organisms like fungus, bacteria, nematode as well as insects. For one sq. m of nursery area, well rotten FYM, 20 g of each N, P and K fertilizer, 2.5 g carbofuran, 200 g of neem cake and 10–25 g tricoderma were applied in the nursery bed.

After sowing the seeds, mulch with green leaves were covered on the bed and irrigation with a rose-can daily in the morning was done. Immediately after germination of the seeds mulch was removed. The seeds were sown on 5th November, 2010.

14.3.6 HARVESTING OF TOMATOES

After proper field preparation, tomato was transplanted on 7th December, 2010. Intercultural operations were done as per standard practices. There are five pluckings which gave the total yield. The first plucking was done on 3rd February 2011, which followed by 15th and 25th February and other two pluckings were done on 10th and 22nd march, 2011.

14.3.7 MATERIAL COMPONENTS FOR DRIP IRRIGATION SYSTEM

14.3.7.1 Pump

In general if the water source is at higher elevation than crop field, water flows under gravity otherwise a pump is used to lift water from the source to deliver it to pipeline system. Capacity and discharge are principal factors to select a pump. Pump capacity is determined on basis of crop water requirement system efficiency and area to be irrigated in a given time. Similarly discharge is determined based on desired operating pressure, functional head loss and change of elevation within the field. Centrifugal pumps are commonly used and deep well turbine pumps in few cases. A centrifugal mono block pump of 3 HP was used to supply water to the drip network in this research work. The source of power for pumping was electricity.

14.3.7.2 Main Line

Poly vinyl chloride (PVC) main pipeline conveys water from the pump to the submain lines and laterals. Pipeline was buried in the soil at suitable depths to avoid hampering tillage operations. Pipe diameter was determined basing on water travel distance and pressure requirement of the system. All filters, bypass valves, fertigation unit, pressure gage unit, etc., were fitted to the main line. The PVC pipe with external diameter of 50 mm with pressure rating 6 kg/cm² was selected in the present study for the mainline supply of the drip system.

14.3.7.3 Submain Line

PVC submain lines supply water from mainline to the laterals on one/both sides of it. Like main line, these were also buried in the soil. Pipe diameter was selected on the basis of the total lateral length and crop water requirement. Ball valves wee provided at head end of submain lines to maintain required pressure and flow in pipe and flush valves at tail end for cleaning of main/submain line. The PVC pipe, external diameter of 40 mm with pressure rating 6 kg/cm^2 was selected on the basis of design to be used as the submain line supply of the drip system.

14.3.7.4 Laterals

Flexible linear low density poly ethylene (LLDPE) lateral pipes were used to convey water from submain line to the emitters. They were laid one/both sides of the submain line and their spacing was decided on basis of row-to-row distance of crop. The laterals are not affected by chemicals/fertilizers/saline water and light in with. One end stop was provided for each lateral at its tail end. During installation of lateral pipes, two types of arrangements were used to study the head drops and discharge variations (Figures 14.2 and 14.3). These are described in the layout procedure. Six laterals were used in both the type of arrangements.

14.3.7.5 Emitters

Drippers/emitters/ticklers/drip nozzles are the most vital component of drip irrigation system which receives water from laterals and delivers to the plant base. Drippers have very slow discharge rate of 4 to 14 lph. The operating pressure at emitter should be always within 1 to 1.5 kg/cm^2. Spacing of drippers on the laterals depends on the type of soil/crop and crop water requirement. Drippers are made of injected hard plastics materials. Drippers having a discharge capacity of 4 lph are selected in the present study for both single and double inlet systems.

14.3.7.6 Filters

Almost 95–98% of the impurities are filtered by the sand filter. Still some minute sand particles and small impurities remain with water for which a screen filter of screen strainer size 20–200 mesh is used. The larger the strainer size, the smaller the particle size that can be filtered out. The maximum flow rate through the screen is 200 gpm/ft^2. A filter of capacity 7m^3/h was used in this research work.

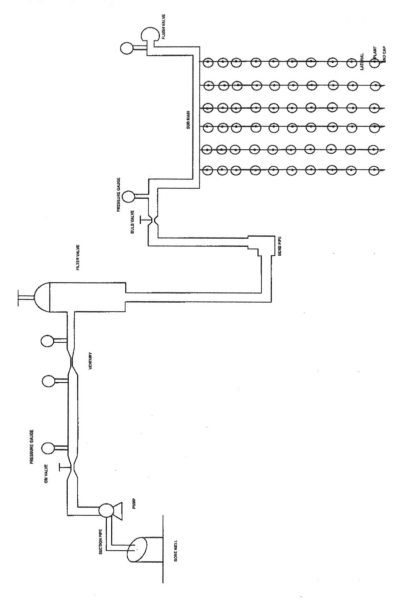

FIGURE 14.2 Layout plan of single inlet drip irrigation system.

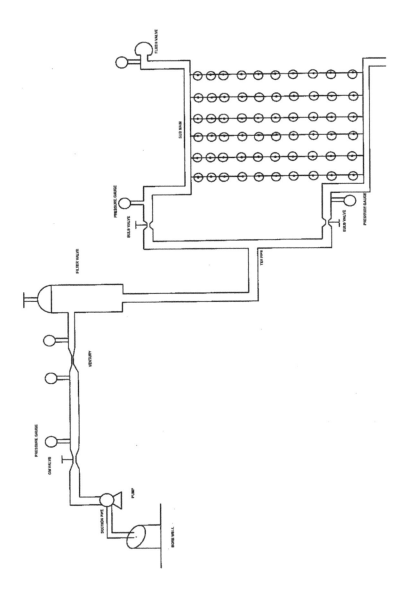

FIGURE 14.3 Layout plan of double inlet drip irrigation system.

14.3.7.7   Bulb Valves

Bulb valves are used to control the flow inside the submain line to maintain pressure having compact double union (DU) design. It is manufactured from high performance rigid PVC compound. It is designed for added safety. Accidental loosening problem of the ball is solved by replacing another one. Chemical resistant nitrile rubber-o-rings ensure leak proof operation for longer period. These are easy to install and dismantle.

14.3.7.8   Flush Valves

Flush valves are used to clean the submain line from any type of debris or impurities which. This is used to maintain the system free from choking for long life. It is fitted at tail end of submain line.

14.3.7.9   Venturi

Drip and micro irrigation have a characteristic not shared by other irrigation methods: fertigation is not optional, but is actually necessary. Fertigation provides the only good way to apply fertilizers physically to the crop root zone. On high value drip irrigated crops, such as lettuce, tomatoes, and peppers, the level of fertigation management for achieving high yields and crop qualities exceeds to what is found with other irrigation methods and crops.

14.3.8   *LAYOUT FOR DRIP IRRIGATION SYSTEM*

As required in case of a drip system, the pipelines and accessories are installed in proper manner. Depending on the row-to-row spacing of tomato crop, the lateral pipes were laid out in the field with a spacing of 60 cm from lateral to lateral (Figures 14.2 and 14.3). Similarly drippers are fixed at a distance of 60 cm depending on the plant-to-plant spacing. As each lateral line is of 6 m in length, therefore 10 drippers were connected to supply water to 10 plants in a row. The length of the submain line was 6 m in both systems and main line was 15 m.

14.3.8.1 Single Inlet System

To study the head drop and discharge variations, two types of arrangements were used in layout of lateral pipes. The length of laterals and spacing of dripper remain unchanged in both the systems. In this arrangement of drip irrigation unit, one end of the lateral line is connected with the submain line and the other end remains plugged (Figure 14.2).

14.3.8.2 Double Inlet System

In this system, both ends of lateral line were connected to two different submain lines in opposite direction, i.e., no end of the lateral line is plugged. The schematic diagram of the layout is given in Figure 14.3.

14.3.9 MEASUREMENT OF DISCHARGE

For carrying out the discharge measurements, time of operation was selected and kept fixed for all lateral lines in both the single and double inlet type arrangements. A stop watch was used to record the time of operation to collect the water discharged from the dripper face, and a properly calibrated measuring glass beaker was used to measure the volume. Discharge measurements were taken at the 1st dripper (entry point) and the last dripper (exit point) of all the six lateral lines of single inlet system under a selected pressure. The numerical differences between these two discharges were tabulated for each lateral line. The whole process of measurement was repeated for the double inlet system.

Then the pressure of supply of water was changed to next value on step by step basis and the discharges were measured for both the single inlet & double inlet systems in liters per hour. The variations in the measurement process were tabulated to draw the conclusions. A drip system usually operates within a pressure range of 1 kg/cm² to maximum 1.5 kg/cm², therefore the values of pressures were kept within this range for carrying out the discharge measurements.

14.3.10   MEASUREMENT OF PRESSURE HEAD

Like discharge measurements, study was also carried out to measure the pressure heads at the drippers at entry and exit points for determination of the pressure drop due to friction in the pipeline in single/double inlet portable drip units. In first phase, applying a particular known pressure to the system, the pressure heads at entry and at exit drippers of each lateral were measured and the differences in values were noted in single inlet drip unit and then in double inlet drip unit. This process was repeated for next values of applied pressure.

In the second phase, both the single and double inlet drip units were considered at entry point and exit point of the lateral lines. That means the, the pressure heads were measured at entry drippers of single inlet and double inlet units and variation in the values were calculated. Similarly pressure heads were measured at the exit point drippers of both the system to find out the deviations. This process was carried out for all values of applied pressure. Pressure head was measured at the dripper ends by help of a pressure gauge in kg/cm² and then the values were transferred to equivalent pressure heads in meters.

14.3.11   STEPS IN DESIGN OF DRIP IRRIGATION SYSTEM

Step-1: Calculation of peak water requirement (PWR)

Under a drip irrigation system, the water requirement includes the crop water demand to meet the losses due to evapotranspiration required for plants to grow. It depends upon the type, stage, effective root zone of crop, soil type, and season of year. The PWR of the crop is calculated as follows:

PWR = (Pan evaporation x pan coefficient
    × crop factor x canopy factor x area)/(Irrigation efficiency)   (1)

Daily pan evaporation was considered in this study. Values of crop coefficient of the crop at different growth stages were taken from FAO publication 24 [26]. The crop factor is indicative parameters for the crop growth and its value depends on foliage characteristics, stage of growth, climatic conditions, etc., Normally 0.7–0.8 is taken for fruit crops and 0.9–1.1 for closely spaced crops (vegetables), etc., The term canopy implies the area of shadow of a plant at 12 noon, which varies according to growth and remains constant after full plant growth. Area of the plant was taken as 0.6 m × 0.6 m which is 0.36 m². In drip irrigation case the irrigation efficiency assumed was 0.90.

Step-2: Selection of dripper
Depending upon the type of dripper and discharge required the number of drippers used can be calculated as:

Number of drippers per plant = Total water required (l/hr) by plant/
Dripper discharge (l/hr)   (2)

Total no. of drippers per row = No. of plants in a row
× No. of drippers/plant   (3)

Dripper discharge may be 4 and 8 lph, taken in general

Step-3: Selection and design of laterals
The length, pipe size and frictional head loss for the laterals are determined by Hazen- William formula. Pressure variation of 5–20% and discharge variation 5–10% throughout the system should be maintained in a range. The pressure difference between first and last dripper should not exceed 10% of normal operating head. Maximum frictional head loss in a lateral should not exceed 2 m in 100 m length of pipe.

Step-4: Selection and design of submain
The hydraulics of submain pipeline is same as that of lateral line. The design includes the determination of submain length and pipe diameter to be used frictional head loss is calculated by help of by using Hazen-William formula. The frictional head loss is limited to be with in the design tolerance of the particular emitter device provided by the manufacturer

Step-5: Selection and design of main

Mainline is designed in a similar way to the submain line. Economy in size, permissible flow velocity, quantity of water flow through it, length of main line, ground elevation, etc., are the factors to be considered for design. In the present study, the head loss in lateral pipe was measured by pressure gage are fitted at particular places where pressure required to be measured.

Step-6: Selection and design of filter

This is based on the factor like-water source, type/size/concentration of physical impurities, quantity of water flowing, ease for handling/cleaning/repairing/economy in investment/power cost, etc., Filters of different filtration capacities (m^3/h) are available for use. In the present research work, total water flowing through main line is 0.13 m^3/h. A screen filter of capacity 3 m^3/h is used in the drip system. Since the water source is a bore well, sand filter was not used.

Step-7: Design of pump

Power is required to lift the required quantity of water from the source and to develop sufficient pressure to operate emitter effectively. The power of the pump can be estimated on the basing of the total head and discharge required to operate system.

$$H = H_e + H_f + H_s + H_E \qquad (4)$$

where, H_e = operating pressure of non-pressure compensating emitters; H_f = HEAD loss due to friction in main/submain as calculated and head loss in lateral is taken from observed data + frictional head loss in filter (observed data) + frictional head loss in fittings assumed + head loss in operating venturi (observed data); H_s = total static head (static suction + static delivery) which is 6 m observed data; H_E = zero for flat land; otherwise the elevation difference between the ground level near the water source and the highest ground level to be irrigated is taken in to account. It is assumed as zero in this study.

The operating pressure is derived from the observed data of Table 14.3 (mentioned later on in this chapter) and is the maximum pressure at which the system operates. Head loss in main and submain pipe is computed as follows:

$$Head\ loss\ in\ main\ and\ submain = (f \times l \times v^2)/(1 \times g \times d) \qquad (5)$$

where, f = friction factor = 0.005; l = length of the pipe (m) which are taken as 15 m and 6 m for main and submain, respectively; v = velocity of water in pipe (m/sec); g = acceleration due to gravity, 9.8 m/sec²; and d = diameter of pipe, (m).

Velocity of water in pipe is calculated from the discharge and diameter of pipe of main and submain are 50 mm and 40 mm, respectively. Frictional head loss in fittings is assumed as 2 m. Now Horsepower required for the pump is calculated below:

$$HP = (H \times Q)/(75 \times a \times b) \qquad (6)$$

where, Q = required discharge in main line (lps); H = total head (m); a = efficiency of motor (assumed 85 %); and b = efficiency of pump (assumed 80%).

14.3.12 YIELD DATA

Yield data of the crop is the summation of yield from all pluckings. In the present experiment, 5 plucking were done for both single inlet and double inlet systems.

14.4 RESULTS AND DISCUSSION

14.4.1 COMPUTATION OF PEAK WATER REQUIREMENT

For installation of a drip irrigation system in the field for any crop, the first and foremost attention should be given to the computation of peak water requirement based on evaporation, stage of crop growth, climatic conditions, irrigation system, efficiency, etc. Such type of assessment of crop water requirement actually helps in deciding the proper pump size and dripper selection for the selected crop. Peak water requirement was also calculated.

There is no meteorological station near the experimental site to record the daily or seasonal evaporation and climatic data. As the area is located within a short distance from Bhubaneswar, so the average daily pan evaporation data for the investigation period (December–March) were collected from the meteorological station of Orissa University of Agriculture and Technology (OUAT), Bhubaneswar and used in computation of peak water requirement. Since peak water requirement is to be estimated, the highest pan evaporation data at peak growth stage of the crop was considered (which is the month of February).

The pan coefficient was assumed as 0.8. The crop factor at mid-season stage, which is maximum, was taken as 1.1.

Similarly the peak water requirement of the crop refers to the stage when the canopy coverage is maximum. In the present experimental study, canopy factor was assumed as 1.0 in computation of peak water requirement. The row-to-row spacing and plant-to-plant spacing of the crop were taken as 0.60 m each. Hence the planting area was 0.36 m². Irrigation efficiency for drip was assumed as 0.9. Using these values, the peak water requirement of the crops was calculated as 1.97 liters/day/plant in winter season (crop growing season).

Therefore, PWR = 1.97 liters/day /plant

Total water requirement for the experimental field consisting of 60 tomato plants = 60 × 1.97=118.27 liters/day

14.4.2 MEASUREMENT OF DISCHARGE

Discharge flow consisted of flow from all the lateral lines (from 1st to 6th lateral pipeline of single inlet system and double inlet system) under several varying pressures. The pre-selected pressure value can be checked or verified by help of pressure gage just at the point of exit of submain line or point of entry to lateral line. Drip system generally operated adequately under an operating pressure range of 1.00–1.50 kg/cm². Therefore four operating pressures were selected for the present experiment: 1.0, 1.1, 1.3 and 1.5 kg/ cm², respectively.

For measuring the discharge, a measuring beaker was taken and the discharge was collected from the beginning of the first lateral pipeline under operating pressure of 1.0 kg/cm^2 within a particular time period measured with the help of stop watch. The discharge rate of dripper was computed by dividing the volume of water collected in the period of observation. Then for the same period of time, discharge was collected at the last dripper of the same 1st lateral line and discharge rate was computed. These two computed values are referred as the 'inlet discharge rate' and 'outlet discharge rate' of lateral number one of the single inlet drip unit, respectively. Thus discharge rates were computed for the rest five of laterals in the unit. Similar process was carried out in the double inlet drip unit for 6 lateral lines to find out the discharge rate. This completed the 1st phase of measurement of discharge rate in the single and double inlet drip systems. In the next step, the operating pressure was carefully changed to next experimental value (1.1, 1.3 and 1.5 kg/ cm^2) respectively. In each case the inlet and outlet discharge rates were measured in both single inlet and double inlet drip units. Tables 14.1 and 14.2 indicate the whole set of observations.

The results obtained in experiments were also utilized to produce graphical form of representation (Figures 14.4–14.11). In Figures 14.4 and 14.5 for operating pressure of 1 kg/cm^2, smooth curves were plotted lateral number versus discharge rate. The values at outlet drippers of 6 lateral lines of double inlet drip unit also gave a similar curve but little above the curve for inlets of laterals. This indicates clearly that under similar physical setup conditions, the dripper discharge at entry point of lateral line is slightly higher in case of double inlet drip unit than those of single inlet drip unit.

Similarly the curves showing the dripper discharge at both inlet and outlet points of lateral lines in single inlet and double inlet drip arrangements were drawn for other operating pressure and it is found that discharge in double inlet system is slightly higher or almost similar to single inlet system (Figures 4.6–4.11).

14.4.3 MEASUREMENT OF PRESSURE HEAD

The pressure measurements were carried out during the course of study in a similar procedure as that of discharge measurement. The whole pro-

TABLE 14.1 Discharge Variations at Inlet and Outlet of Single Inlet and Double Inlet Systems

Lateral No.	Single inlet			Double inlet		
	Inlet discharge	Outlet discharge	Variation	Inlet discharge	Outlet discharge	Variation
	lph					
Pressure 1 kg/cm²						
1	3.86	3.74	0.12	3.89	3.77	0.12
2	3.85	3.73	0.12	3.88	3.76	0.12
3	3.82	3.71	0.11	3.87	3.75	0.12
4	3.81	3.70	0.11	3.86	3.74	0.12
5	3.79	3.68	0.11	3.85	3.73	0.12
6	3.78	3.67	0.11	3.81	3.70	0.11
Pressure 1.1 kg/cm²						
1	3.87	3.75	0.12	3.87	3.75	0.12
2	3.86	3.75	0.12	3.87	3.75	0.12
3	3.85	3.73	0.12	3.86	3.74	0.12
4	3.85	3.73	0.12	3.86	3.74	0.12
5	3.84	3.72	0.12	3.86	3.74	0.12
6	3.83	3.71	0.11	3.85	3.74	0.12
Pressure 1.3 kg/cm²						
1	3.98	3.86	0.12	3.99	3.87	0.12
2	3.87	3.75	0.12	3.90	3.78	0.12
3	3.86	3.74	0.12	3.90	3.78	0.12
4	3.85	3.73	0.12	3.89	3.77	0.12
5	3.84	3.72	0.12	3.89	3.77	0.12
6	3.82	3.71	0.11	3.89	3.77	0.12
Pressure 1.5 kg/cm²						
1	3.99	3.87	0.12	4.00	3.88	0.12
2	3.99	3.87	0.12	4.00	3.88	0.12
3	3.98	3.86	0.12	3.99	3.87	0.12
4	3.98	3.86	0.12	3.99	3.87	0.12
5	3.98	3.86	0.12	3.99	3.87	0.12
6	3.98	3.86	0.12	3.98	3.86	0.12

TABLE 14.2 Discharge Variations as Affected by Varying Pressures for Single Inlet and Double Inlet Systems

Lateral No.	Inlet Discharge, lph				Outlet Discharge, lph			
	Single inlet	Double Inlet	Devia-tion	Varia-tion, %	Single inlet	Double Inlet	Devia-tion	Varia-tion, %
Pressure, 1 kg/cm²								
1	3.86	3.89	0.03	0.78	3.74	3.77	0.03	0.78
2	3.85	3.88	0.03	0.78	3.73	3.76	0.03	0.78
3	3.82	3.87	0.05	1.31	3.71	3.75	0.05	1.31
4	3.81	3.86	0.05	1.31	3.70	3.74	0.05	1.31
5	3.79	3.85	0.06	1.58	3.68	3.73	0.06	1.58
6	3.78	3.81	0.03	0.79	3.67	3.70	0.03	0.79
Pressure, 1.1 kg/cm²								
1	3.87	3.87	0.00	0.00	3.75	3.75	0.00	0.00
2	3.86	3.87	0.01	0.21	3.75	3.75	0.01	0.21
3	3.85	3.86	0.01	0.26	3.73	3.74	0.01	0.26
4	3.85	3.86	0.01	0.26	3.73	3.74	0.01	0.26
5	3.84	3.86	0.02	0.52	3.72	3.74	0.02	0.52
6	3.83	3.85	0.02	0.65	3.71	3.74	0.02	0.65
Pressure, 1.3 kg/cm²								
1	3.98	3.99	0.01	0.25	3.86	3.87	0.01	0.25
2	3.87	3.90	0.03	0.67	3.75	3.78	0.03	0.67
3	3.86	3.90	0.04	0.93	3.74	3.78	0.03	0.93
4	3.85	3.89	0.04	1.09	3.73	3.77	0.04	1.09
5	3.84	3.89	0.05	1.41	3.72	3.77	0.05	1.41
6	3.82	3.89	0.07	1.73	3.71	3.77	0.06	1.73
Pressure, 1.5 kg/cm²								
1	3.99	4.00	0.01	0.20	3.87	3.88	0.01	0.20
2	3.99	4.00	0.00	0.08	3.87	3.88	0.00	0.08
3	3.98	3.99	0.01	0.18	3.86	3.87	0.01	0.17
4	3.98	3.99	0.01	0.18	3.86	3.87	0.01	0.18
5	3.98	3.99	0.01	0.18	3.86	3.87	0.01	0.18
6	3.98	3.98	0.00	0.10	3.86	3.86	0.00	0.10

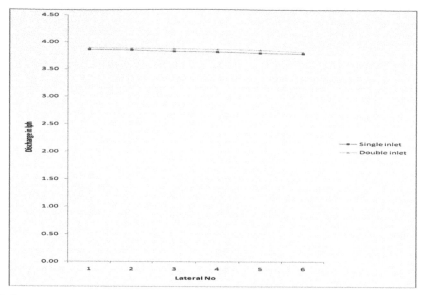

FIGURE 14.4   Inlet discharge curve of single and double inlet system (pressure = 1.0 kg/cm²).

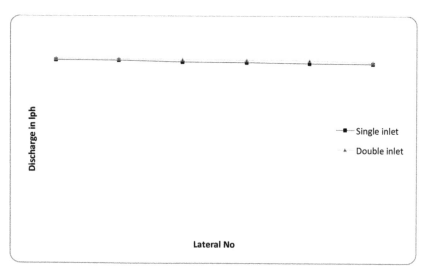

FIGURE 14.5   Outlet discharge curve of single and double inlet system (pressure = 1.0 kg/cm²).

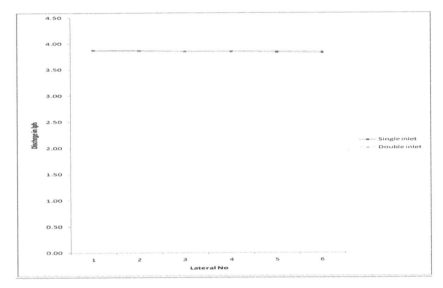

FIGURE 14.6   Inlet discharge curve of single and double inlet system (pressure = 1.1 kg/cm²).

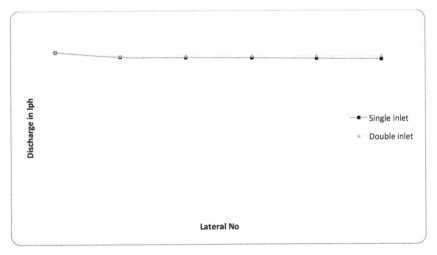

FIGURE 14.7   Outlet discharge curve of single and double inlet system (pressure = 1.1 kg/cm²).

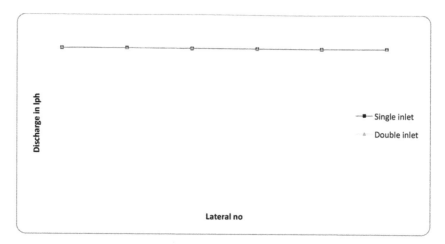

FIGURE 14.8   Inlet discharge curve of single and double inlet system (pressure = 1.3 kg/cm²).

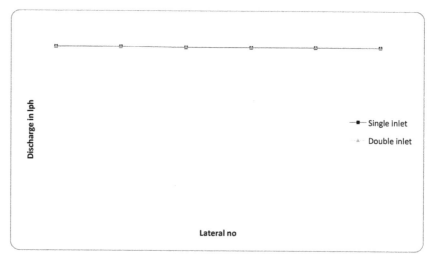

FIGURE 14.9   Outlet discharge curve of single and double inlet system (pressure = 1.3 kg/cm²).

cess was divided into two sets of observations. In the 1ˢᵗ phase of measurement, the operating pressure for the drip system was selected as 1 kg/cm² which was kept unchanged for both the single inlet and double inlet drip units. Under this working pressure, water was allowed to flow

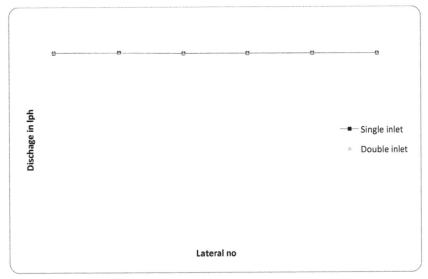

FIGURE 14.10 Inlet discharge curve of single and double inlet system (pressure = 1.5 kg/cm²).

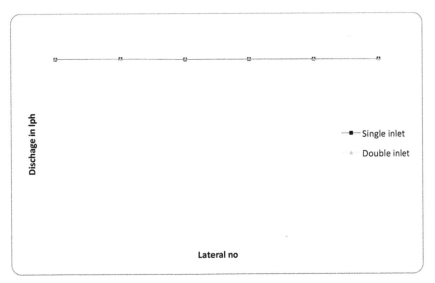

FIGURE 14.11 Outlet discharge curve of single and double inlet system (pressure = 1.5 kg/cm²).

in the system. The pressure heads were measured at the points just before the 1st (entry) dripper and at the exit dripper to find out the inlet and outlet pressures.

To accomplish this, the lateral line was cut at a point just before the first and the last dripper, again joined by two poly 'T' to which the pressure gage were attached to take the readings. The readings were converted to equivalent pressure heads (m). After the measurements were taken, the gages were removed and the open ends of the 'T' were closed by help of end plugs. This process was carried out for six lateral lines and deviations in pressure heads between inlet and outlet points were calculated for single inlet drip unit and then for double inlet drip unit. In case of single inlet unit, the outlet point (exit dripper) was at midpoint of lateral line which was valid for both the submain lines, the entry point dripper being considered from either of two submain lines. Operating pressure were changed from 1.0 kg/cm² to 1.1, 1.3 and 1.5 kg/cm², respectively, to find out other readings and the whole set of observation are given in Table 14.3.

14.4.4   COMPARISON OF PRESSURE LOSSES

Pressure loss due to friction is directly proportional to the length of pipe, velocity of flow but is inversely proportional to the diameter of pipe. In the present investigation, the diameter of submain and lateral pipes were not changed, thus velocity of flow was also constant. The change in parameter is only the effective length of lateral pipe.

The effective length of lateral pipe is 6 m in case of single inlet drip arrangement and in case of double inlet drip arrangement the effective length is 3 m. Obviously the pressure drop due to friction in the pipe is more in case of single inlet unit than that in double inlet unit. The same have been reflected in Tables 14.3 and 14.4. These values were plotted in Figures 14.12–14.19 and the curves were similar in all cases. As observed at operating pressure of 1 kg/cm², pressure head loss variation was 0.08 to 0.9 m in case of single inlet unit, compared to 0.02 m in case of double inlet unit. For other operating pressures, the variations in pressure head were within similar range as these values.

TABLE 14.3 Pressure Head Variation (m) at Inlet and Outlet of Single Inlet and Double Inlet Systems

Lateral No.	Single inlet, m			Double inlet, m		
	Inlet pressure	Outlet pressure	Variation	Inlet pressure	Outlet pressure	Variation
Pressure, 1.0 kg/cm²						
1	8.60	8.51	0.09	10.00	9.98	0.02
2	8.59	8.51	0.08	9.99	9.97	0.02
3	8.55	8.46	0.09	9.94	9.92	0.02
4	8.46	8.38	0.08	9.84	9.82	0.02
5	8.39	8.30	0.09	9.75	9.73	0.02
6	8.20	8.12	0.08	9.54	9.52	0.02
Pressure, 1.1 kg/cm²						
1	9.46	9.37	0.09	11.00	10.97	0.03
2	9.32	9.22	0.1	10.83	10.80	0.03
3	9.37	9.28	0.09	10.89	10.86	0.03
4	9.32	9.22	0.1	10.83	10.80	0.03
5	9.25	9.15	0.1	10.75	10.72	0.03
6	9.21	9.12	0.09	10.71	10.68	0.03
Pressure, 1.3 kg/cm²						
1	11.18	11.07	0.11	13.00	12.965	0.035
2	11.12	11.01	0.11	12.93	12.895	0.035
3	11.05	10.94	0.11	12.85	12.815	0.035
4	11.02	10.91	0.11	12.81	12.775	0.035
5	11.01	10.9	0.11	12.8	12.765	0.035
6	11.00	10.89	0.11	12.79	12.755	0.035
Pressure, 1.5 kg/cm²						
1	12.90	12.77	0.13	15.00	14.962	0.038
2	12.85	12.72	0.13	14.94	14.902	0.038
3	12.77	12.64	0.13	14.85	14.812	0.038
4	12.73	12.6	0.13	14.8	14.762	0.038
5	12.69	12.56	0.13	14.75	14.712	0.038
6	12.60	12.47	0.13	14.65	14.612	0.038

TABLE 14.4 Effect of Pressure on Pressure Head of Both Single and Double Inlet Systems

Lateral No.	Inlet pressure head in m				Outlet pressure head in m			
	Single inlet	Double Inlet	Devia-tion	Varia-tion, %	Single inlet	Double Inlet	Devi-ation	Varia-tion, %
Pressure, 1.0 kg/cm²								
1	8.60	10.00	1.40	16.28	8.51	9.98	1.47	17.27
2	8.59	9.99	1.40	16.30	8.51	9.97	1.46	17.16
3	8.55	9.94	1.39	16.26	8.46	9.92	1.46	17.26
4	8.46	9.84	1.38	16.31	8.38	9.82	1.44	17.18
5	8.39	9.75	1.36	16.21	8.3	9.73	1.43	17.23
6	8.20	9.54	1.34	16.34	8.12	9.52	1.40	17.24
Pressure, 1.1 kg/cm²								
1	9.46	11.00	1.54	16.28	9.37	10.97	1.60	17.08
2	9.32	10.83	1.51	16.20	9.22	10.80	1.58	17.14
3	9.37	10.89	1.52	16.22	9.28	10.86	1.58	17.03
4	9.32	10.83	1.51	16.20	9.22	10.80	1.58	17.14
5	9.25	10.75	1.50	16.22	9.15	10.72	1.57	17.16
6	9.21	10.71	1.50	16.29	9.12	10.68	1.56	17.11
Pressure, 1.3 kg/cm²								
1	11.18	13.00	1.82	16.28	11.07	12.97	1.90	17.12
2	11.12	12.93	1.81	16.28	11.01	12.90	1.89	17.12
3	11.05	12.85	1.80	16.29	10.94	12.82	1.88	17.14
4	11.02	12.81	1.79	16.24	10.91	12.78	1.87	17.09
5	11.01	12.80	1.79	16.26	10.90	12.77	1.87	17.11
6	11.00	12.79	1.79	16.27	10.89	12.76	1.87	17.13
Pressure, 1.5 kg/cm²								
1	12.90	15.00	2.10	16.28	12.77	14.96	2.19	17.17
2	12.85	14.94	2.09	16.26	12.72	14.90	2.18	17.15
3	12.77	14.85	2.08	16.29	12.64	14.81	2.17	17.18
4	12.73	14.80	2.07	16.26	12.60	14.76	2.16	17.16
5	12.69	14.75	2.06	16.23	12.56	14.71	2.15	17.13
6	12.60	14.65	2.05	16.27	12.47	14.61	2.14	17.18

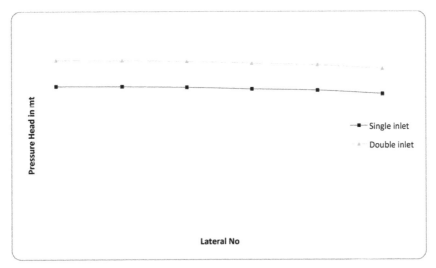

FIGURE 14.12   Inlet pressure curve of single and double inlet system (pressure = 1 kg/cm²).

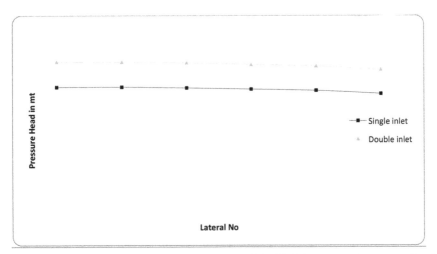

FIGURE 14.13   Outlet pressure curve of single and double inlet system (pressure = 1 kg/cm²).

When operating pressure is 1 kg/cm², the pressure head at inlet point of the D.I. unit was found within 9.54 to 10.00 m (0.95 to 1 kg/cm²). It is almost same as the operating pressure indicating that there is no head drop due to pipe friction. But in case of S.I. unit, the pressure head was

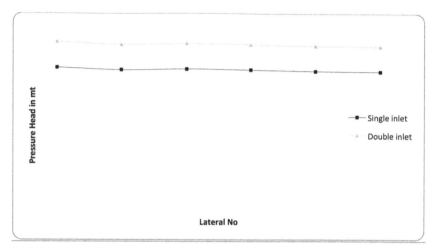

FIGURE 14.14   Inlet pressure curve of single and double inlet system (pressure = 1.1 kg/cm²).

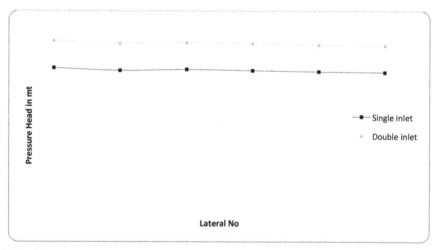

FIGURE 14.15   Outlet pressure curve of single and double inlet system (pressure = 1.1 kg/cm²).

within 8.20 to 8.60 m (0.82 to 0.86 kg/cm²), which indicates a drop in pressure due to pipe friction. Similar results were observed in case of pressure heads measured at outlet point of the S.I., & D.I. units. Even by increasing the operating pressure from 1.0 kg/cm² to 1.1, 1.3, 1.5 kg/cm² etc., pressure drops did not change significantly.

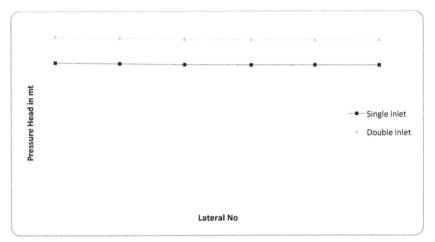

FIGURE 14.16   Inlet pressure curve of single and double inlet system (pressure = 1.3 kg/cm²).

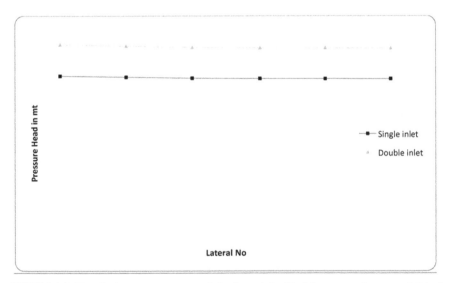

FIGURE 14.17   Outlet pressure curve of single and double inlet system (pressure 1.3 kg/cm²).

14.4.5   ESTIMATION OF PUMP CAPACITY

Generally capacity of the pump depends upon the total flow (Q) of the mainline and total head (H) occurs during operation of the system.

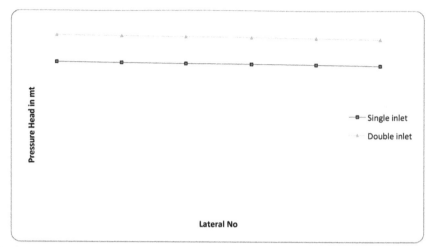

FIGURE 14.18 Inlet pressure curve of single and double inlet system (pressure = 1.5 kg/cm²).

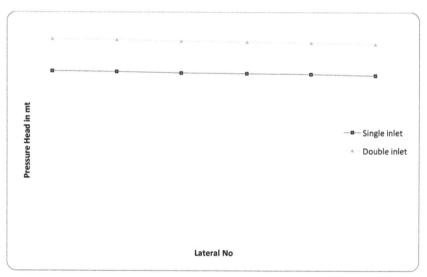

FIGURE 14.19 Outlet pressure curve of single and double inlet system (pressure = 1.5 kg/cm²).

Total flow, Q = (Total number of plants × Dripper discharge)/3600 = (60 × 4)/3600 = 0.066 lps

Total head loss for single inlet system, H

Suction and delivery head loss = 6 m (observed data)

Venturi head loss = 5 m

Filter head loss = 5 m

Fittings head loss = 2 m

Operating pressure = 15 m

Head loss in main and submain line = 0.17 and 0.06 m (calculated)

Head loss in laterals = 0.78 m (observed data)

Hence total head loss, H = 6 + 5 + 5 +2 + 15 + 0.17 + 0.06 + 0.78 = 34.01 m

Capacity of pump = [0.066 × 34.01]/constant = 0.044 HP

Total head loss for double inlet system, H

Suction and delivery head loss = 6 m (observed data)

Venturi head loss = 5 m

Filter head loss = 5 m

Fittings head loss = 2 m

Operating pressure = 15 m

Head loss in main and submain line = 0.17 and 0.06m (calculated)

Head loss in laterals = 0.23 m (observed data)

Hence total head loss, H = 6 + 5 + 5 +2 + 15 + 0.17 + 0.06 + 0.23 = 33.68 m

Capacity of pump = [0.066 x 33.68]/constant = 0.0435 HP

Savings in HP = [(0.044 – 0.0435)] × 100/0.044 = 1.13 %

14.4.6 TOMATO YIELD

In the present study, yield from five plucking were summed to calculate the yield variation in both single and double inlet system (Table 14.5). It is observed that there is 9% increase in yield in double inlet drip irrigation system as compared to the single inlet drip irrigation system. Hence, it is recommended that farmers may adopt double inlet drip irrigation system than the conventional single inlet drip irrigation system in their field, since there is saving of pump capacity and increase in yield of tomato.

TABLE 14.5 Yield (kg per experiment area) in Single Inlet and Double Inlet System

Harvesting	Single inlet	Double inlet
	Tomato yield, kg/plot	
1st	10	12
2nd	13.5	14.5
3rd	25	26
4th	20	22.5
5th	9.5	10
Total	78	85

14.5 CONCLUSIONS

From the study in this chapter, following conclusions can be drawn:

- In double inlet drip irrigation, 16–17% head loss is minimized as compared to single inlet system.
- The flow of discharge in double inlet system is slightly more and uniform as compared to single inlet system.
- Due to the uniform flow, nutrient distribution also occurs uniformly at the root zone of the plant which gives more yield than single inlet system.
- Due to less head loss occurs in double inlet system, the pump required to operate the system is of less capacity, giving a saving of 1.13%. It will be more for large and compact patch.
- Due to higher yield and less pump capacity, the productivity of the crop will be increased in double inlet system.
- In double inlet system, the cost of drip is slightly increased and it is replenished by its net return.

14.6 SUMMARY

Drip irrigation is now gaining accelerating popularity in supplying precise and economical amount of irrigation especially to vegetables and orchard crops. But the installation and accessory costs are too high for the resources that cause some problems in adoption. There is a strong need to carry out research so that the cost of accessories especially the pump and pipe costs can be minimized. To reduce the pump cost, head loss in the system need to be minimized.

The present study was undertaken with two types of drip irrigation systems, i.e., (i) single inlet system and (ii) double inlet system. In single inlet system, laterals are connected to submains in one side only. In double inlet system, two submains are used at two ends and laterals are connected to both the submains at two ends. The study revealed that head loss and pressure drop is lower across the laterals in double inlet drip system. The uniformity in discharge is hence more than the single inlet drip system. The pump capacity required in double inlet system was obtained to be less as compared to single inlet system. Further, there is about 9% increase in yield of tomato in case of double inlet drip irrigation system. Therefore. Authors suggest that the farmers should adopt the double inlet drip irrigation system than the conventional single inlet drip irrigation system in vegetable crops.

KEYWORDS

- bulb valves
- canopy factor
- crop factor
- delivery head
- discharge
- double inlet system
- drip irrigation
- dripper
- efficiency
- experiment
- filter
- flush valves
- friction coefficient
- head loss

- **horse power**
- **irrigation**
- **irrigation efficiency**
- **lateral**
- **main pipe**
- **micro irrigation**
- **motor**
- **pan coefficient**
- **pan evaporation**
- **peak water requirement**
- **pressure**
- **pump**
- **single inlet system**
- **suction head**
- **tomato**
- **venturi**
- **yield**

REFERENCES

1. Ahmed, I. (1994). Significance of energy losses due to emitter connections in trickle irrigation lines. *J. Agric. Eng. Res*, *60*, 1–5.
2. Bagarello, V., Ferro, V., Provenzano, G., & Pumo, D. (1995). Experimental study on flow resistance law for small-diameter plastic pipes. *J. Irrig. Drain. Eng.*, *121*(5), 313–316.
3. Christiansen, J. E. (1942). *Irrigation by sprinkling*. California Agric. Exp. Stn. Bull, 670, University of California, Berkeley.
4. Dandy, G. C., & Hassanli, A. M. (1996). Optimum design and operation of multiple subunit drip irrigation systems. *J. Irrig. Drain. Eng.*, *122*(5), 262–275.
5. Giuseppe, P., & Domenico, P. (2004). Simplified procedure to evaluate head losses in drip irrigation laterals. *Journal of Irrigation and Drainage Engineering*, *131*(6), 318–324.
6. Hughes, T. C., & Jeppson, R. W. (1978). Hydraulic friction loss in small diameter plastic pipelines. *Water Resour. Bull.*, *14*(5), 1159–1166.
7. John, D. (2003). Explicit hydraulic design of micro irrigation submain units with tapered manifold and laterals. *Journal of Irrigation and Drainage Engineering*, *123*(2), 227–236.

8. Juana, L., & Alberto, L. (2002). Discussion of determining minor head losses in drip irrigation laterals. *Journal of Irrigation and Drainage Engineering, 128*(6), 376–384.

9. Juana, L. E., Rodrıguez, G., & Alberto, L. (2002). Minor head losses in drip irrigation laterals. *Journal of Irrigation and Drainage Engineering, 121*(3), 376–384.

10. Kamand, F. Z. (1988). Hydraulic friction factors for pipe flow. *Journal of Irrigation and Drainage Engineering, 114*(2), 311–323.

11. Liou, C. P. (1988). Limitations and proper use of the Hazen-Williams equation. *J. Hyrdaul. Eng., 124*(9), 951–954.

12. Maisiri, N., Sanzanje, A., Rockstrom, J., & Twomlow, S. J. (2005). On farm evaluation of the effect of low cost drip irrigation on water and crop productivity compared to conventional surface irrigation system. *Phys. Chem. Earth, 30*, 783–791.

13. Moody, L. F. (1944). Friction factors for pipe flow. *Transactions of the ASME, 66*, 120–144.

14. Nayak, S. C. (2007). Energy drops by friction in portable drip irrigation units. *Journal of Research (Orissa University of Agricultural and Technology), 25*(2), 139–141.

15. Pedro, V., Pedro, L., & Luque, E. (2001). New algorithm for hydraulic calculation IN irrigation laterals. *Journal of Irrigation and Drainage Engineering, 127*(3), 254–260.

16. Provenzano, G., Pumo, D., & Dio, P. (2005). Simplified procedure to evaluate head losses in drip irrigation laterals. *Journal of Irrigation and Drainage Engineering, 131*, 527–530.

17. Ravindra, V. K., Singh, R. P., & Mahar, P. S. (2008). Optimal design of pressurized irrigation subunit. *Journal of Irrigation and Drainage Engineering, 134*(2), 137–146.

18. Reddy, K. (2004). Evaluation of on-line trickle irrigation emitter barb losses. *Journal of Institution of Engineers (India), Division of Agricultural Engineering*, 42–46.

19. Romero, P., Garcia, G., & Botia, P. (2006). Cost-benefit analysis of a regulated deficit-irrigated almond orchad under subsurface drip irrigation conditions in southeastern Spain. *Irrig. Sci., 24*, 175–184.

20. Sarbu, I. (1997). Optimal design of water distribution networks. *Journal of Hydraulic Research, 35*(1), 63–79.

21. Swamee, P. K., & Jain, A. K. (1976). Explicit equations for pipe flow problems. *J. Hydr. Div., 102*(5), 657–664.

22. Terzidis, G. (1992). Discussion of simple and accurate friction loss equation for plastic pipe. *J. Irrig. Drain. Eng., 118*(3), 501–504.

23. Tiwari, K. N., & Reddy, K. (1997). Economic analysis of trickle irrigation system considering planting geometry. *Agricultural Water Management, 34*, 195–206.

24. Von Berunath, R. D. (1990). Simple and accurate friction loss equation for plastic pipe. *J. Irrig. Drain. Eng., 116*(2), 294–298.

25. Von Berunath, R. D., & Wilson, T. (1989). Friction factors for small diameter plastic pipes. *J. Hydraul. Eng., 115*(2), 183–192.

26. Wu, I. (1991). Energy line approach for direct hydraulic calculation in drip irrigation system. *Irrig. Sci, 13*, 21–29.

CHAPTER 15

FERTIGATION IN A DRIP IRRIGATION SYSTEM: EVALUATION OF VENTURI INJECTORS AND ITS SIMULATION STUDY

S. V. CHAVAN, B. S. POLISGOWDAR, A. B. JOSHI, M. S. AYYANAGOWDER, U. SATISHKUMAR, and V. B. WALI

CONTENTS

15.1 INTRODUCTION

In micro irrigation, fertilizers can be applied through the system with the irrigation water directly to the region where most of the plants roots develop. This process is called fertigation and it is done with the aid of special fertilizer apparatus (injectors) that is installed at the head control unit of the system, before the filter. Fertigation is a necessity in drip irrigation, though not in the other micro irrigation installations, although it is highly recommended and easily performed

Fertigation not only saves fertilizer and man – hours, but also improves fertilizer use efficiency and reduce production costs. Fertilization device is a one of the main device in the drip irrigation system. At present, commonly applied devices for fertilization includes pressure difference fertilizer applicator, fertilizer pump and venturi injector.

Venturi operates on the principle of vacuum suction created by an advanced venturi complex. This implements the latest know-how in hydraulic technology and allows the injectors to operate at low-pressure differential. A vacuum is created as the water flows through a converging passage that gradually widens. Injection is activated when there is a pressure differential between the water entering the injectors and the water with chemical leaving into the irrigation system. The injection rate of venturi using fresh irrigation water has been reported to be directly proportional to the pressure drop across it [1].

This research study was conducted to test the different fertigation equipments (Venturi injector, domestic fertigation and fertilizer tank) to evaluate the hydraulic performance of the system [3].

15.2   MATERIALS AND METHODS

The experiment was conducted at Jain Plastic Park, Jain Irrigation Systems Ltd., Jalgaon (Maharashtra). The Jain Plastic Park lies between 75°32'55" E to 75°33'20" E longitude and 21°00'05" N to 21°00'20" N latitude. It is about 227 m above mean sea level. The climate of the area is semi-arid with 800 mm mean annual rainfall. Laboratory test was carried out at the Jain Plastic Park. The company Jain Irrigation Systems Ltd. started in 1963 is one of the pioneers of micro irrigation systems in India.

The details of the experimental setup for 63 mm diameter venturi are shown in Figure 15.1. Experimental set up for venturi injector was designed by carefully considering the venturi motive flow rate, types of venturi injector, maximum and minimum pressure, and motive flow rate through venturi and pump capacity.

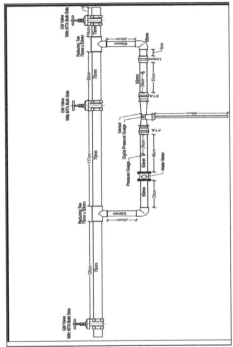

FIGURE 15.1 Experimental setup for 63 mm φ venturi injector.

The pipes in the study were all made of polyvinyl chloride (PVC). The size of the main diameter of the pipe was 75 mm. The pressure gages were installed 240 mm away from the venturi injector and the flow meter was installed 460 mm away from the venturi injector. The GM$_2$ valve was installed between the two reducing tees and 1200 mm away from the inlet PVC reducing tee. GM$_3$ valve was installed 140 mm away from the outlet PVC reducing tee. Both inlet and outlet pressures could be changed by adjusting these two valves (GM$_2$ and GM$_3$).

The tests were conducted for a constant inlet pressure and varying outlet pressures and under Indian farm conditions. The experiment and CFD analyzes for 63 mm Venturi injector were conducted for inlet pressure of 2.0 kg cm^{-2} and outlet pressure of 0.8, 1.0, 1.2 and 1.4 kg cm^{-2}.

All experimental data were recorded after both inlet and outlet pressures were stabilized after 3 minutes. The injection rate, motive flow rate and throat pressure were measured each time as the injection was started at different pressure drop. The structural details of four venturi injectors having 63 mm inlet diameter are presented Table 15.1.

TABLE 15.1 Structural Design of the Commercially Available 63 mm φ Venturi Injectors having the Same Inlet and Outlet diameters

Particulars	Type "A"	Type "B"	Type "C"	Type "D"
Convergent angle	21° 04'	21° 44'	17° 6'	19° 55'
Convergent length (mm)	29.90	27.99	30.14	43.93
Diameter of throat opening (mm)	4.95	3.90	5.50	17.60
Diameter of the throat (mm)	19.21	17.95	18.00	19.60
Divergent angle	3° 56'	5° 09'	2° 90'	3°34'
Divergent length (mm)	144.00	144.09	193.00	172.00
Inlet diameter (mm)	40.00	40.00	40.00	40.00
Length of the throat (mm)	26.10	20.00	40.00	29.85
Outlet diameter (mm)	40.00	40.00	40.00	40.00

15.2.1 MEASUREMENT OF THE MOTIVE FLOW RATE

Actual motive flow (liter) was measured during the experiment by using the flow meter. Theoretical motive flow was calculated by using Bernoulli's equation [4]:

$$Q_{the} = {}^{A_2}\sqrt{[2\,(P_1 - P_2)]/\,[\rho\,\{1 - (B)^4\}]} \tag{1}$$

$$A_2 = (\pi/4) \times d^2 \tag{1a}$$

where, B = throat diameter and inlet diameter in m; d = throat diameter in m; P_1 = inlet pressure in kg cm^{-2}; P_2 = throat pressure in kg cm^{-2}; ρ = density of water; A_2 = cross sectional area of throat in m^2.

15.3 COMPUTATIONAL FLUID DYNAMICS (CFD)

CFD model is used to develop certain framework for the implementation of knowledge-based applications to support the design of products requiring complex virtual and experimental analysis. Characteristic input and output parameters are used for the specific problem and test cases in the knowledge base play the role of the experiments.

Experiment has been conducted on hydraulic study, design and analysis of different geometries of drip irrigation emitter labyrinth [4] using CFD model to predict motive flow rate through emitter and analyze its hydraulic performance under various water pressures.

15.3.1 THEORETICAL CONSIDERATIONS IN CFD MODEL

The brief description of various components and the mathematical relationships used to simulate the processes and their interactions in the CFD model are described below. Water flow in the Venturi injector can be regarded as incompressible, steady flow, which is consistent with laws of mass and momentum conservation. Therefore, the basic equations are:

Continuity equation

$$\frac{\partial u}{\partial x} + \frac{\partial v}{\partial y} + \frac{\partial w}{\partial z} = 0 \tag{2}$$

Navier-Stokes equation

$$\frac{\partial(\rho u)}{\partial t} + \nabla (\rho u U) = \frac{\partial p}{\partial x} + \mu \nabla^2 u + F_x \tag{3}$$

$$\frac{\partial(\rho v)}{\partial t} + \nabla (\rho u U) = \frac{\partial p}{\partial y} + \mu \nabla^2 v + F_y \tag{4}$$

$$\frac{\partial(\rho w)}{\partial t} + \nabla (\rho u U) = \frac{\partial p}{\partial z} + \mu \nabla^2 w + F_z \tag{5}$$

where, u (m.s^{-1}) velocity in × direction; v (m.s^{-1}) velocity in y direction; w (m.s^{-1}) velocity at z direction; t (s) is time; U (m.s^{-1}) is velocity vector; ρ(kg m^{-3}) is the water density; p (Pa) is pressure of fluid tiny body; μ (Pa·s) is dynamic viscosity coefficient of water; Fx, Fy, Fz are force components on unit volume at x, y, z directions, respectively.

15.3.2 DATA BASE FOR CFD ANALYSIS

The data were collected from the field based on the type of Venturi injector. The information from the field includes: type of Venturi injector, size of Venturi injector, inlet and outlet pressure, pump capacity, area under fertigation, type of crop under irrigation. The inlet and outlet pressures collected from field survey were used as input in the CFD analysis. The main intension of survey was to prepare database required for CFD analysis. The study of Venturi injector was conducted through the survey to gather the information for Venturi CFD analysis.

15.3.3 INPUT PARAMETERS

The input parameters can be classified as geometrical (structural parameters of Venturi injector), physical (Inlet and outlet pressure, Fluid and wall conditions), process and operating (Boundary conditions, goals, faces, unit system). The input data required for CFD analysis are presented in Table 15.2.

15.3.4 COEFFICIENT OF DISCHARGE

- Experimental analysis: It is the ratio of the actual discharge to the theoretical discharge:

TABLE 15.2   Input Data Required for CFD Analysis

Input data	Description
Structural models	To support the design of products requiring complex virtual and experimental analysis
Boundary conditions	A boundary condition is required where fluid enters or exits the model and can be specified as an inlet pressure, outlet pressure and throat pressure
Goals	Goals are set to determine pressure drop and flow rate through Venturi injector
Faces	Faces on which conditions and goals are specified
Analysis type	Internal; Exclude cavities without flow conditions
Fluid	Water
Unit system	SI system
Wall Conditions	Zero roughness

$$C_d(\text{Expt.}) = \frac{Q_{act}}{Q_{the}} \tag{6}$$

- CFD analysis: It is the ratio of the actual discharge calculated by the computational fluid dynamics (CFD) analysis to the theoretical discharge[6]:

$$C_d(\text{CFD}) = \frac{Q_{cfd}}{Q_{the}} \tag{7}$$

15.4 RESULTS AND DISCUSSION

The performance of the Venturi injector depends on the pressure drop, structural parameters of the Venturi and motive flow through Venturi injector.

15.4.1 PRESSURE AVAILABILITY AT INLET SIDE ACCORDING TO FIELD SURVEY

The main intension of survey was to collect all possible information of Venturi injector at field level which was an input data to work on CFD model and also type of methodology followed by farmers for fertigation.

The inlet pressure varied in the range of 0.9 to 2.5 kg cm^{-2}. It was observed that the maximum inlet pressure was recorded at a 2.0 kg cm^{-2} (Figure 15.2). Inlet pressure available at the inlet of Venturi injector at most of the sites was about 2.0 kg.cm^{-2}. This pressure data was an input for CFD analysis to estimate injection rate, motive flow at variable pressure at outlet (i.e., 0.8, 1.0, 1.2, 1.4 kg cm^{-2}).

The survey was conducted at different irrigation sites, where Venturi injector has been installed in Sangli and Kolhapur districts in Maharashtra. The data collected from the fields was divided based on the type of Venturi injector 63 mm. Information from the field was: type of Venturi injector, size of Venturi injector, inlet and outlet pressure, pump capacity, area under fertigation, type of crop under irrigation. The inlet and outlet pressures collected from field survey were used as input in the CFD analysis.

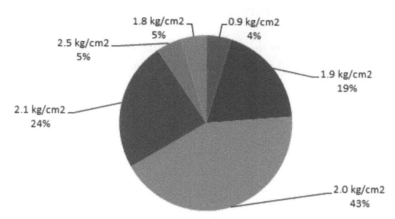

FIGURE 15.2   Pressure availability at inlet side as per field survey for 63 mm Venturi injector.

15.4.2 COMPARISON OF THE 63 MM VENTURI INJECTOR BY CFD AND EXPERIMENTAL ANALYSIS

Table 15.3 shows the relationship between the outlet pressures, injection rate and motive flow for 2kg.cm^{-2} inlet pressure. From the results of experimental analysis, maximum injection rate of 32.18 lpm with motive flow of 302.34 lpm was observed in TYPE "B" Venturi injector when the outlet pressure was 0.8 kg cm^{-2} (i.e., at the pressure drop of 1.2 kg cm^{-2}). TYPE "B" Venturi injector was observed to be giving maximum injection rate for each differential pressure. Injection rate of 9.35 lpm with motive flow 276.26 lpm was observed in TYPE "B" Venturi injector when the outlet pressure was 1.4 kg cm^{-2} (i.e., at the pressure drop of 0.6 kg cm^{-2}). At the same outlet pressure, other types of Venturi injectors failed to inject water.

The results of the CFD analysis are presented in Table 14.3 and depicted in Figures 15.3–15.6. The Figure 15.7 presents the pictorial view of experimental set up and collection of data.

From the CFD analysis results in TYPE "B" Venturi injector were observed to give maximum injection rate of 35.10, 33.26, 18.12, and 12.65 lpm with a motive flow of 304.25, 302.89, 276.12 and 272.23 lpm, respectively, for outlet pressure of 0.8, 1.0, 1.2 and 1.4 kg cm^{-2} (Figure 15.4).

TABLE 15.3 Comparison of the 63 mm Venturi Injector by CFD Analysis and Experimental Values

Type of Venturi injector	Flow (lpm)	0.8 kg cm⁻² Outlet pressure		1.0 kg cm⁻² Outlet pressure		1.2 kg cm⁻² Outlet pressure		1.4 kg cm⁻² Outlet pressure		df	χ^2 cal. Value at 1% sig.	χ^2 tab. Value at 1% sig.	Results
		Expt.	CFD	Expt.	CFD	Expt.	CFD	Expt.	CFD				
TYPE "A"	MF	321.72	330.23	297.01	325.61	295.49	300.48	NIL	NIL	2	2.80	9.21	NS
	IR	28.60	34.25	23.89	30.54	15.58	17.47	NIL	NIL	2	2.57	9.21	NS
TYPE "B"	MF	302.47	304.25	301.49	302.89	281.31	276.12	276.26	272.23	3	0.17	11.3	NS
	IR	32.18	35.10	28.31	33.26	17.88	18.12	9.35	12.65	3	1.05	11.3	NS
TYPE "C"	MF	118.00	122.13	116.07	120.38	112.44	115.54	NIL	NIL	2	0.36	9.21	NS
	IR	8.75	7.23	6.58	5.89	3.28	3.59	NIL	NIL	2	0.42	9.21	NS
TYPE "D"	MF	129.63	133.65	123.51	128.65	121.56	116.85	NIL	NIL	2	0.51	9.21	NS
	IR	11.09	12.35	9.16	10.89	5.64	7.98	NIL	NIL	2	1.08	9.21	NS

MF = Motive flow; IR = Injection rate.

The CFD analysis results (Table 15.3) indicated that TYPE "B" Venturi injector was observed to give injection rate 12.65 lpm with motive flow 272.23 lpm when the outlet pressure was 1.4 kg cm^{-2} (i.e., at the pressure drop of 0.6 kg cm^{-2}). At the same outlet pressure, other types of Venturi injector failed to inject water.

It has been also observed from the CFD analysis results that TYPE "C" Venturi injector was having minimum injection rate of 7.23, 5.89, and 3.59 lpm with a motive flow of 122.13, 120.38 and 115.54 lpm, respectively, for an outlet pressure of 0.8, 1.0 and 1.2 kg cm^{-2}. It was observed that the TYPE "B" Venturi injector could inject maximum water for pressure drop ranging from 0.6 up to 1.5 kg cm^{-2}. Similar findings have also been reported by Nesthad et al. [3].

It was observed that structural parameters of the TYPE "B" Venturi injector having convergent angle 21°44′, divergent angle 5°09′, convergent length 27.99 mm, throat diameter 17.95 mm, diameter of throat opening 3.90 mm and length of the throat 20 mm are more effective parameters for maximum injection rate with minimum motive flow than the other types of Venturi injectors. The divergent angle of the TYPE "B" Venturi injector (5°09′) was observed being the highest than the other types of Venturi injector, which avoided the possibility of flow separation and the consequent energy loss [2].

In TYPE "B" Venturi injector, the acceleration of the flowing liquid allowed to take place rapidly in a relatively small convergent length (27.99 mm), without resulting appreciable loss of energy. The divergent length of 144.09 mm with divergence 5°09′ was observed in TYPE "B" Venturi injector, which avoided turbulence in liquid. Similar findings were reported by Xinga et al. [6].

The statistical analysis showed that there were no significant differences among the experimental values and CFD analysis values. It can be observed that both experimental and CFD data have exactly similar trends.

15.4.3   COMPARISON CD VALUES OF VENTURI INJECTOR BY CFD AND EXPERIMENTAL VALUES

Coefficient of discharge (C_d) was evaluated by using the results from CFD analysis and was compared with experimental C_d values. Table 15.4

TABLE 15.4 Comparison of Coefficient of Discharge of 63 mm Venturi Injector by CFD Analysis and Experimental Study

Type of Venturi injector	Coefficient of discharge, C_d								df	χ^2 cal. Value at 1% sig.	χ^2 tab. Value at 1% sig.	Results
	At outlet pressure, kg cm^{-2}											
	0.8		1.0		1.2		1.4					
	Expt.	CFD	Expt.	CFD	Expt.	CFD	Expt.	CFD				
TYPE "A"	0.69	0.72	0.66	0.73	0.64	0.68	NI	NI	2	0.01	9.21	NS
TYPE "B"	0.81	0.82	0.80	0.81	0.77	0.76	0.76	0.76	3	0.000037	11.3	NS
TYPE "C"	0.33	0.35	0.32	0.34	0.31	0.33	NI	NI	2	0.000353	9.21	NS
TYPE "D"	0.36	0.37	0.34	0.36	0.34	0.34	NI	NI	2	0.000138	9.21	NS

df = degrees of freedom, NS = non-significant, C_d = coefficient of discharge, χ^2 = Chi-square.

showed in both cases that an increase in the outlet pressure decreases the C_d value. It was seen from the Table 15.4 that among all outlet pressures, the maximum C_d was observed in TYPE "B" and the next maximum was observed in TYPE "A" Venturi injector.

Statistically the CFD simulation results showed close proximity to the experimental results. The Table 15.4 revealed that in 63 mm Venturi injector maximum coefficient of discharge was observed in TYPE "B" Venturi injector for both CFD and experimental analysis. The maximum C_d value in experimental analysis and CFD analysis were 0.81, 0.80, 0.77, 0.76 and 0.82, 0.81, 0.76, 0.76, respectively for outlet pressure of 0.8, 1.0, 1.2 and 1.4 kg cm^{-2}. These results are in close agreement with the findings obtained by Tamhankar et al. [5].

15.5   CONCLUSIONS

As per the Indian farm conditions, the CFD analysis for 63 mm Venturi injectors were conducted for inlet pressure of 2 kg cm^{-2} and outlet pressure of 0.8, 1.0, 1.2 and 1.4 kg cm^{-2}. The CFD results showed that TYPE "B" Venturi injector could inject the maximum water with minimum motive flow than the other types of Venturi injectors.

The chi-square test results showed that there were no significant differences among the predicted (CFD model) and observed (experimental) values. Both CFD model and experimental values have similar trend in injection rate, motive flow and coefficient of discharge. The results illustrate that in both analyzes the maximum injection rate with minimum motive flow was observed in TYPE "B" Venturi injector. To optimize the structural parameters of Venturi injector which can give maximum injection rate with minimum motive flow, conducting experiments would involve lot of money and the time. From the present study, it is concluded that laboratory study can be replaced with CFD analysis to optimize the structural parameters of Venturi injectors.

Based on the present study, the 63 mm Venturi injector having desired structural parameter for maximum injection rate with minimum motive flow as per the operating conditions prevailing in the field will having following dimensions:

Inlet diameter = 40.00 mm
Convergent angle = 21°44′
Divergent angle = 5°09′
Convergent length = 27.99 mm
Divergent length = 144.00 mm
Diameter of the throat = 17.95 mm
Diameter of the throat opening = 3.90 mm
Length of the throat = 20.00 mm
Outlet diameter = 40 m

15.6 SUMMARY

In this study an attempt was made to study the performance of 63 mm size venture injector for fertigation and drip irrigation system. Four different sizes of Venturi by manufacturers were considered for the study. Computational fluid dynamics model was selected for flow simulation of venture injector. The inlet pressure was maintained at 2 kg cm^{-2} whereas outlet pressures were 0.8, 1.0, 1.2 and 1.4 kg cm^{-2}. Type B venture injector was found to give best results in respect of in respect of various performance evaluation parameters.

KEYWORDS

- actual discharge
- chi-square test
- coefficient of discharge
- computational fluid dynamics
- continuity equation
- convergent angle
- divergent angle
- drip irrigation
- emitter
- fertigation

- **flow parameters**
- **injection rate**
- **inlet pressure**
- **manufacturer**
- **micro irrigation**
- **motive flow**
- **Navier-Stoke equation**
- **outlet pressure**
- **simulation**
- **structural parameter**
- **theoretical discharge**
- **throat diameter**
- **venturi injector**

REFERENCES

1. Kumar, M., Rajput, T. B. S., & Patel, N. (2013). Performance evaluation of commercially available Venturi under inlet and outlet pressure differentials. *J. Agric. Engg.*, *42*(4), 11–13.
2. Modi, P., & Seth, S. (2013). Hydraulics and Fluid Mechanics Including Hydraulic Machines. *Standard Book House*, Delhi.
3. Nesthad, N., Kurien, E. K., Varughese, A., & Mathew, E. K. (2013). Evaluation of different fertigation equipments and the hydraulic performance of the drip fertigation system. *J. Agric. Sci.*, *1*(1), 12–17.
4. Patil, S. S., Nimbalkar, P. T., & Joshi, A. B. (2013). Hydraulic study, design and analysis of different geometries of drip irrigation emitter labyrinth. *Int. J. Eng. Advanced Tech.*, *2*, 455–462.
5. Tamhankar, N., Pandhare, A., & Bansode, V. (2014). Experimental and CFD analysis of flow through venturi meter to determine the coefficient of discharge. *Int. J. Eng. and Tech.*, *3*, 194–200.
6. Xinga, H., & Miao, W. (2009). CFD simulation to the flow field of venturi injector. *J. Computer and Computing Technologies in Agric.*, *2*, 805–815.

CHAPTER 16

CANAL WATER BASED PRESSURIZED IRRIGATION SYSTEMS IN HIGH RAINFALL AREAS

R. C. SRIVASTAVA and S. MOHANTY

CONTENTS

16.1   INTRODUCTION

India has one of the largest and most ambitious irrigation program in the world with net irrigated area exceeding 47 million hectares. However, the overall project efficiency of the surface irrigation system in India is very low, which leads to poor utilization of irrigation potential created at huge cost. The efficiency of canal irrigation system in India has been reported to vary between 30–35% [7]. Low project efficiency not only results in poor utilization of irrigation potential created at huge cost, but also aggravates the degradation of soil and water resources. Therefore the need of the hour

In this chapter: 1.00 US$ = 60.00 Rs. (Indian Rupees).

is to increase the irrigation efficiency of existing projects and use saved water for irrigating new areas.

At present average overall project efficiency of several canal irrigation projects in the rice growing areas in the world has been estimated to be 23% and that of non-paddy crops to be 40% [11]. International Institute for Land Reclamation and Improvement (ILRI), Netherlands reviewed the conveyance losses in irrigation supply schemes of different countries of the world and reported maximum conveyance loss of 60% in India and minimum in Philippines (13%) [1]. In other countries like Austria, USA, Spain, Columbia, Egypt, Greece and Italy, the conveyance losses range from 40 to 59% and in Japan, Australia, South Korea, Malaysia, Taiwan, France, the losses range from 16 to 37%. About 71% of the irrigation water is lost in the whole process of its conveyance from head works and application in the field. The break-up of the losses consists of main and branch canal (15%), distributaries (7%), water courses (22%) and field losses of 27% [6].

The situation is particularly worse in minor irrigation systems of plateau areas of eastern India, where the overall irrigation efficiency varies from 20 to 35%. These systems are located in coarse soil regions with rolling topography. Due to this, the conveyance losses are high and the system suffers from inadequate supply and poor water availability especially during lean season. Therefore, the need of the hour is to increase irrigation efficiency of existing projects and use saved water for irrigating new areas or reducing the gap between potential and actual irrigated areas.

Shifting to pressurized irrigation systems can be an option for increasing this irrigation efficiency. It has been reported that average application efficiency of 14 sprinkler irrigated projects was 70%, for 15 basin and wild flooding projects was 45% and for 24 other and combined method projects was 59% [11]. It has also been reported that on-farm irrigation efficiency for trickle irrigation can theoretically approach 90 to 95% [2]. The efficiency for surge irrigation and LEPA system has been reported to be as high as 80% and 95% respectively [3]. A shift in application method from surface to pressurized system has potential of vastly improving irrigation application efficiency. However, the design and operation of this system in a canal command will be different from a typical tube well based project.

The plateau areas of eastern India are characterized by high rainfall ranging between 1100 to 1600 mm. As significant rainfall occurs during monsoon season (June–October), only rice crop can be grown in medium and low lands, which form major chunk of canal command area. The canal water carries a heavy silt load especially during monsoon, which has to be taken care of before this water is used for pressurized irrigation. The land holdings in the command of an outlet are small with large number of owners of varied socio economic background and therefore the command area has diversified cropping pattern especially during post monsoon season, each requiring different irrigation method. Further the area has potential of aquaculture, which can be beneficially exploited while changing the supply from on-off to continuous.

Based on these constraints and potential, a canal-based pressurized irrigation system should satisfy following conditions:

- It should have an adjunct service reservoir to maintain continuous supply of water.
- It should provide surface irrigation to rice in monsoon season and pressurized irrigation to non-rice crops during dry season.
- The system should be integrated to operate different types of systems, viz., sprinkler, drip and micro-sprinkler to suit various crops under one outlet.
- It should have provision for removal of silt deposited in the pipe network during operation for surface irrigation.
- It should have provision of aquaculture for increasing production by non-consumptive utilization of water body.

This chapter presents a case study to find out the feasibility of shifting from surface irrigation to pressurized irrigation system in a canal irrigation system.

16.2   CASE STUDY

A study was carried out at ICAR-Indian Institute of Water Management (IIWM) Research Farm, Deras, Bhubaneswar (Odisha), India to find out the feasibility of shifting from surface irrigation to Pressurized irrigation system in a canal irrigation system. The ICAR-IIWM research farm lies in the command of Deras Minor Irrigation System. The study was conducted during the period 2001–2005.

16.2.1 CONSTRUCTION OF ADJUNCT RESERVOIR

To maintain continuous supply of water, one adjunct reservoir of 2500 m³ was constructed to store the diverted water from the field channel serving to the other part of research farm as well as farmers. The capacity of the service reservoir was estimated using the formula [8]:

$$V = 10.A.I.n + (SP + 0.7\ Eo).a.n/1000 + 0.75\ a \qquad (1)$$

where, V = volume of tank in m³, A = command area in ha, I = weighted gross irrigation demand in mm/day, n = off period of canal in days, SP = seepage loss in mm/day, Eo = open pan evaporation in mm/day, a = surface area of the tank in m².

In this study, the values were: A=5 ha, I = 5 mm/day (considering irrigation efficiency), n = 7 days, SP = 4 mm/day, Eo = 5 mm/day and a = 900 m². Using Eq. (1) with these values, the capacity of the pond was found out approximately as 2500 m³.

16.2.2 INSTALLATION OF PRESSURIZED IRRIGATION SYSTEM

A filtration unit was installed along with the pump so that sediments in the water can be removed to prevent emitter clogging in the drip irrigation system. The filtration unit was a three-stage filtration process involving a hydrocyclone filter, a sand filter and a screen filter. A catch well was constructed between the adjunct reservoir and the pumping-cum-filtration unit to facilitate the pumping of water. A hybrid sprinkler and drip system was installed with 110 mm PVC pipeline as the mainline for both sprinkler and drip system. The diameter of the pipeline was 110 mm so that the friction loss in pipe is less than the gravity head available due to slope. The planning was done in such a way so that irrigation can be done to the paddy crops by gravity flow during monsoon season through pipe conveyance and surface irrigation. In the post-monsoon season, pumping based sprinkler and drip irrigation system were used to irrigate the crops by pumping water from reservoir through catch well. The 10 HP electrical pump was installed for this purpose. Under sprinkler irrigation, crops like pea, potato, French beans, cowpea and sunflower were cultivated. Under drip irrigation, crops like tomato, maize, okra, marigold and capsicum were cultivated.

The system was designed in such a way that both sprinkler and drip irrigation sets can be operated simultaneously. Four outlets were taken out from the mainline to irrigate 2.8 ha area by sprinkler irrigation system and two outlets were taken out to irrigate 1.9 ha by drip irrigation. In the drip irrigation system, 4 lph and 2 lph pressure compensating drippers were used on a paired row basis. Maize, Okra and tomato were irrigated by 4 lph drippers whereas marigold and capsicum were irrigated by 2 lph drippers. As the sprinkler irrigation system requires a head of about 2 kg/cm^2 and drip irrigation system requires a head of 1 kg/cm^2, the first four outlets were used for sprinkler irrigation and next two outlets were used for drip irrigation system. The available pressure in the sprinkler irrigation system varied from 2 to 2.5 kg/cm^2 and in the drip irrigation system, it varied from 1.5 to 2 kg/cm^2. The service reservoir was designed with two outlets at different elevations in such a way so that the water flow from upper outlet by gravity during monsoon for surface irrigation and it flows through the other outlet to a catch well for pumping during post monsoon season. The schematic diagram of this canal based pressurized irrigation system is shown in Figure 16.1. The view of the adjunct reservoir along with the pump house and on-dyke horticultural plants is shown in Figure 16.2. The view of potato crop irrigated by sprinkler irrigation in the command area is shown in Figure 16.2.

16.2.3 EVALUATION OF IRRIGATION EFFICIENCY

To assess the efficiency of the pressurized irrigation system, the irrigation efficiency of this system was evaluated in comparison to surface irrigation system in another part of command of the same outlet with similar cropping system. The parameters evaluated were conveyance efficiency, application efficiency, and uniformity of the pressurized system. The standard procedure for estimating these parameters were adopted [4]. The conveyance efficiency of earthen conveyance channel (200 m length) was measured by RBC flume. The discharge was measured at the source and at the delivery point to the field and the conveyance efficiency was found to be 75.07%. Uniformity coefficient of the sprinkler irrigation system along the periphery and along the radius was found to be 82.61% and 80.2%, respectively. The uniformity coefficient values were compared with the corresponding wind speed values and it was found that the uniformity coefficient value decreases with wind speed. Emission uniformity of drip irrigation system was evaluated for individual laterals and the whole system. The mean value of emission uniformity for individual laterals and whole system were found to be 97.1 % and 94.2%, respectively. Generally, a uniformity coefficient of 90% in case of drip irrigation and 85% in sprinkler irrigation is considered satisfactory [5]. So, the uniformity coefficient in case of drip irrigation is well within the satisfactory level whereas in case of sprinkler irrigation it is slightly below the satisfactory level.

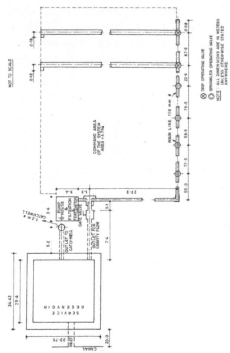

FIGURE 16.1   Schematic diagram of canal based pressurized irrigation system [10].
(Source: Srivastava RC, Ahmed M (1998) Design criteria for canal supply based pressurized irrigation system in high rainfall areas of eastern India. Progress in micro-irrigation research and development in India. In: Proc. nat. seminar on micro-irrigation research in India: status and perspectives for 21st century, 27–28 July, 1998, Bhubaneswar, pp 90–95 and [10]: fig 2 on page 3023I)

The application efficiency of surface, drip and sprinkler irrigation system was estimated on different dates. In surface irrigation system and drip irrigation system, it was evaluated in tomato crop; and in sprinkler irrigation, it was evaluated in cowpea crop. The water stored in the root zone of the crops was divided by the amount of water delivered to the field in the corresponding time period in order to determine the application efficiency. The mean application efficiency of the surface irrigation system, sprinkler irrigation system and drip irrigation systems were estimated and found as 61.47, 77.2, and 90.19%, respectively [9]. Studying this along with conveyance efficiency, it can be said that the irrigation efficiency below the outlet for surface irrigation system was 46.14%, against that of 77.2% for sprinkler and 90.19% for drip (assuming 100% conveyance efficiency). This indicates that the irrigation efficiency increases by 67.32% in case of sprinkler system and 95.47% in case of drip system.

Apart from the hydraulic study of irrigation efficiency, the evaluation of the system was done in terms of crop production, horticultural plants on the embankment and pisciculture in the service reservoir.

16.2.4 CROP PRODUCTION

A rice-based crop rotation was grown to evaluate its performance in terms of the productivity and water use efficiency under surface irrigated condition and pressurized irrigation system. Another area was put under same crop rotation with open channel conveyance and compared with the above system. During monsoon the only difference between these two was that inside project area it was pipe conveyance and surface application while outside project area it was open channel conveyance and surface application.

During dry season, the comparison was between surface irrigation with open channel conveyance and pressurized irrigation. While in monsoon and winter season, the crops were put up under both types of application methods, no crop was planted with surface irrigation during summer as the canal is closed by 1st week of April. During summer (March end to mid June), a part of drip-irrigated command was put under maize and okra, which was irrigated by water stored in the reservoir. It was observed that in monsoon season also, there was saving of water, although there was no

FIGURE 16.2. iew of the adjunct reservoir along with pump house (top); View of sprinkler irrigated potato in the command area (bottom).

significant difference in the yield under both irrigation systems. For post monsoon crops, there was significant jump in yield, water saving and irrigation water productivity for all the crops. Table 16.1 shows that average total annual saving of water in a command area of 4.7 ha was 12610 m³. Net water saving through the system is 11790 m³. Thus shift to the pressurized system increased the irrigation efficiency as well as the productivity.

16.2.5 HORTICULTURAL CROPS

Papaya was planted on the top of the embankment of the reservoir and creeping type cucurbits on its outward slope. The data on crop growth was monitored every month and the fruits were harvested whenever it was ready. It was found that 1025, 1010 and 1645 kg of fruits were obtained from the plants in 2002–03, 2003–04 and 2004–05, respectively. The yield was lower in earlier years, as the soil was excavated earth from the pond and therefore was poor in fertility. Bottle gourd (*Lagenaria vulgaris*) was grown on the outer embankment of the pond and the average annual production was 211 kg.

16.2.6 PISCICULTURE

In the service reservoir, fry of Indian major carps were stocked @ 10,000/ ha along with advanced fingerlings @ 2,000/ha for a period of 8 months.

TABLE 16.1 Water Balance of the System (average of three crop years)

Item	Average amount (m³)
1. Water saved in monsoon	3960
2. Water saved in winter by sprinkler	5960
3. Water saved in *rabi* by drip irrigation	1540
4. Water saved during summer	1150
5. Total gross water saved = 1+2+3+4	12610
6. Water lost from tank through evaporation from October to May	680
7. Water lost by seepage loss @ 2 mm per day from October to May	140
8. Net water saving through the system = (5–6–7)	11,790

The growth of the fish and water quality parameters were monitored for the whole growth period. However, to avoid loss on account of poaching the production process was leased for an annual fee of Rs. 5,650.00 (In this chapter: One US$ = 60.00 Rs., Indian Rupees). Growth performance of *C.mrigala* was higher than that of *L.rohita* probably due to the fact that being bottom dweller, *C.mrigal* is more tolerant to oxygen depletion. After 8 months of rearing, the mean body weights were 885, 460, and 520 g, for *C.catla, L.rohita* and *C.mrigala,* respectively. Faster growth rate was, however, recorded for *C.catla* (3.68 g ADG) followed by *C.mrigala* (2.16 g ADG) and *L.rohita* (1.91 g ADG). The critical water quality variables remained within the optimal range. The average yield was 2.34 t/ha.

16.2.7. ECONOMIC ANALYSIS OF THE SYSTEM

Economic analysis of the system was done by calculating the cost of saving the unit amount of the water by the system. For this, the benefits due to increased production of the crops in the command during monsoon and winter season, additional production from summer crops, horticultural crops on embankment of the adjunct reservoir and pisciculture in the reservoir were estimated. The cost was taken as the annual cost of the system inclusive of reservoir (which include depreciation, interest on investment, maintenance and energy cost for operating the pump), cost of cultivation of summer crops, horticultural crops on embankment and pisciculture. The amount of the water saved was estimated by calculating water saved during monsoon and winter season, water available for the summer cultivation and then deducting the water lost from the adjunct reservoir due to seepage and evaporation. The total income from summer crops has been accounted as benefit as no crop can be grown during summer with existing system because the canal is closed in 1st week of April. The water saved will either irrigate additional area or meet the demand of other sector. In absence of this saving, new water resource will have to be created to meet this demand. In view of it, the cost of saved water was added to the overall return from the system.

Table 16.2 shows that annual additional benefit from crops and allied components was Rs. 55,000.00. Table 16.3 estimates the annual cost of the system by calculating the cost of the adjunct reservoir and drip and sprin-

TABLE 16.2 Net Additional Income from Different Activities

Crop	Area (ha) (2)	Av. yield gain (t/ha) (3)	Net additional production (Kg) (4) = (2 + 3) × 1000	Net additional income (Rs.) —
Capsicum	0.10	1.90	190.0	945
Cowpea	0.18	0.24	43.2	450
French bean	0.20	0.28	56.0	855
Maize	0.25	0.50	125.0	630
Marigold	0.50	0.50	250.0	15,000
Okra	0.40	0.40	160.0	945
Pea	0.28	0.06	16.8	270
Potato	1.70	3.08	5238.0	13,095
Rice	4.50	0.19	855.0	4,275
Sunflower	0.52	0.24	124.8	1,890
Tomato	0.55	4.55	2502.5	5,000
Total additional return from crops =				43,355
Fish				5,355
Papaya and bitter gourd				6,280
Total additional return = 43,355 + 5355 + 6280 =				55,000

kler system separately, which was Rs. 76,300.00. Hence the additional annual receipt from the crops is less than the annual cost of the system. But there is significant water saving in the system. Hence by adding the cost of the saved water with the annual receipt from the crops, annual benefit from the system is obtained which is higher than the annual cost of the system. The benefit–cost ratio of the system was 1.126 [10]. This B:C ratio will improve further if the cultivation during summer season is enlarged and pisciculture in pond is better managed. However, the initial cost of the system is high and cannot be funded by individual farmers. It has to be done either by government, who has social obligation to enlarge irrigated area for maintaining food security or an outside agency, which will utilize the saved water.

Government agencies in India are slowly handing over a part of the network of the canal irrigation systems to *pani panchayats* (water user's associations) for its operation and maintenance. Hence, once the canal based pressurized irrigation system is in place, the 'pani panchayat' can maintain the system with the collection of water tax from the farmer beneficiaries. The construction of auxiliary reservoirs can be fitted into the government program of rehabilitation of canal irrigations systems, which are followed by transfer of irrigation management to 'pani panchayats.'

TABLE 16.3   Annual Cost of the System and Benefit–Cost Analysis

S. No.	Item	Cost (Rs.)
1.	Investment on pond and pump house	200,000
2.	Investment on drip and sprinkler system including pumping system	300,000
Sub total ... A		500,000
Annual cost of Pond and Pump House		
1.	Depreciation on pond and pump house (assuming 25 years life and 50% junk value)	4,000
2.	Maintenance @ 1%	2,000
3.	Interest on investment @ 9%	9,000
Subtotal ... B		15,000
Annual cost of Drip and sprinkler System		
1.	Depreciation on drip and sprinkler system assuming 10 years life and 10% junk value	27,000
2.	Maintenance of the system @ 2%	6,000
3.	Interest on investment @ 9%	13,500
4.	Electricity charges for 500 hours of 10 hp pump	8,300
Subtotal ... C		54,800
Additional cost of cultivation		
1.	Annual cost of fish cultivation and papaya cultivation	1,500
2.	Annual cost of cultivation of Okra and Maize	5,000
Subtotal ... D		6,500
Total annual cost of the system ... E = C + D		76,300
Total additional receipt from crops		55,000
Water saved per annum		11,790 m³
Annual cost of providing 1 ha-m (or 10,000 m³) of water at outlet		26,250
Equivalent cost of water saved		30,950
Annual benefit from the system ... F		85,950
Benefit cost ratio of the system dividing annual benefit by annual cost = **F** ÷ **E** = **G**		1.126

16.3   CONCLUSIONS

It can be concluded that the canal based pressurized irrigation is a feasible option in flow based minor irrigation systems in plateau areas; and it increases irrigation efficiency very significantly. The system reduced the turbidity of the water and provided continuous supply of water so that pressurized irrigation systems can be used with the canal irrigation system. The uniformity coefficient of drip and sprinkler irrigation systems was well within the acceptable limits. The benefit–cost ratio of the system was 1.126. The canal based pressurized irrigation system with adjunct reservoir has the potential of becoming a good way of irrigation in canal command of minor irrigation systems. However, since the initial capital cost is higher, Government has to take initiative in view of social, ecological and economic benefits from the system.

16.4   SUMMARY

The efficiency of canal based irrigation system in India is very low especially in rice fields. On the other hand, the Pressurized irrigation system such as the drip and sprinkler irrigation system have higher efficiency to a tune of about 90–95%. A canal-based pressurized irrigation system should have an adjunct service reservoir to maintain continuous supply of water. It should provide surface irrigation to rice in monsoon season and pressurized irrigation to non-rice crops during dry season. The system should be integrated to operate different types of systems, viz., sprinkler, drip and micro-sprinkler to suit various crops under one outlet.

KEYWORDS

- **adjunct reservoir**
- **annual benefit**
- **annual cost**
- **application efficiency**
- **benefit cost ratio**

- **canal water**
- **command area**
- **conveyance efficiency**
- **drip irrigation**
- **dry season**
- **economic analysis**
- **irrigation efficiency**
- **micro-sprinkler**
- **minor irrigation system**
- **monsoon season**
- **non-rice crop**
- **outlet**
- **Pani panchayat**
- **pisciculture**
- **plateau region**
- **pressurized irrigation**
- **project efficiency**
- **return**
- **rice crop**
- **service reservoir**
- **sprinkler irrigation**
- **surface irrigation**
- **uniformity coefficient**

REFERENCES

1. Bos, M. G., & Nugteren, J. (1990). *On Farm Irrigation Efficiencies*. International Institute for Land Reclamation and Improvement, Netherlands, 21 pp.
2. Bucks, D. A., Nakayama, F. S., & Warrick, A. W. (1982). Principles, Practices and Potentialities of Trickle Irrigation. *Advances in Irrigation. Ed. D. Hillel*, Academic Press, New York. pp. 220–291.
3. Daniel, T. (1995). *TAES Research Focuses on Center Pivot Irrigation Management Decisions.* TAES, 10 pp.

4. Michael, A. M. (1978*). Irrigation Theory and Practice*. Vikas Publications, New Delhi, 339 pp.

5. Murty, V. V. N. (2002). *Land and Water Management Engineering*. Kalyani Publishers, Ludhiana, pages 335.

6. Navalwala, B. N. (1991). Waterlogging and its related issues in India. *J. Irrigation and Power,* 55–64.

7. Sanmuganathan, K., & Bolton, P. (1988). Water management in third world irrigation schemes – Lesson from the field. *ODU Bull., 11, Hydraulic Research*, London, UK, 28 pp.

8. Srivastava, R. C., & Ahmed, M. (1998). Design criteria for canal supply based pressurized irrigation system in high rainfall areas of eastern India. Progress in Micro-Irrigation Research and Development in India. *Proc. Nat. Seminar on Micro-Irrigation Research in India: Status and Perspectives for 21ˢᵗ Century*, 27–28 July, Bhubaneswar, pp. 90–95.

9. Srivastava, R. C., Mohanty, S., Singandhupe, R. B., Biswal, A. K., Ray, L. I. P., & Sahoo, D. (2006). Studies on canal water based pressurized irrigation system in a minor irrigation command. *Journal of Agricultural Engineering, 43*(4), 28–35.

10. Srivastava, R. C., Mohanty, S., Singandhuppe, R. B., Mohanty, R. K., Behera, M. S., Ray, L. I. P., & Sahoo, D. (2010). Feasibility evaluation of pressurized irrigation in canal commands. *Water Resources Management, 24*(12), 3017–3032.

11. Walters, W., & Bos, M. G. (1989). Irrigation Performance, Assessment and Irrigation Efficiency. Annual Report by *International Institute for Land Reclamation and Improvement, ILRI,* Wageningen, Netherlands, 231 pp.

INDEX

Milton Keynes UK
Ingram Content Group UK Ltd.
UKHW050300161024
449569UK00048B/798